Roads and Bridges of the Roman Empire

Horst Barow

Roads and Bridges of the Roman Empire

Edition Axel Menges

Funding for this book was provided by Frauke Barow, who was also a great help in the daily work of Edition Axel Menges in publishing the work of her late husband, Horst Barow.

ISBN 978-3-936681-53-6

Printing and binding: Graspo CZ. a. s., Zlín, Czech Republic

Translation into English and editing: Friedrich Ragette
Typesetting and editing: Dorothea Duwe
Prepress work: Axel Menges, Reinhard Truckenmüller
Design: Axel Menges
Layout: Helga Danz

Foreword

During a journey through Spain I noticed on a road near Mérida the sign »TRANSPORTES ESPECIALES«. Following it, I found an old stone bridge with two semicircular arches. Apparently, the highway administration deemed the modern reinforced concrete bridge not strong enough for very heavy vehicles and rather trusted the solidity of Roman arch construction, built 2000 years ago.

This rather accidental discovery sparked my decision to study Roman bridge building in depth. Based upon more than 30-years civil engineering experience, having been involved in the construction of numerous roads and concrete bridges, I increasingly realized, what technical masterworks the Roman engineers achieved in the space of seven centuries.

However, we need no bridge when there is no road, so I decided to study the Roman highways as well. I also realized from my own working experience, that I would not have been able to build a single road or bridge without a functioning government to commission it, or without a highly developed construction industry to build it.

I therefore decided to organize my study from the top down, beginning with the Roman political and administrative background, the available technology and material means, before entering into *medias res.*

I was fortunate to visit many of the ancient sites, and I consider the case studies of bridges, each with many personal illustrations and observations, an important part of this study.

Horst Barow

1. Introduction

»But the greatest of the physical monuments, which occupied the best energies of Roman surveyors, planners, engineers, labourers, masons and slaves for centuries and made possible the growth and administration of the largest empire the world had hitherto known, was neither a mighty building nor a statue but a thing both ponderously physical and entirely horizontal and thus, at least from a distance, rather hard to see: certainly invisible, and very hard to imagine, as a whole. This was the enormous road system without which the Roman Empire could not have existed. Estimates of its size vary a good deal, depending on how many secondary and tertiary roads are figured in. But it was certainly not less than 80,000 kilometres and possibly as much as 100,000 or even 120,000, including its many bridges thrown over foaming rivers, culverts above swamps and tunnels hewn through mountainous rock. It was a stupendous feat of surveying, planning and labour, and all done without earth-moving machines, graders or explosives – just hand-tools and muscle.«

Robert Hughes, 2011

While there are many studies of Roman fortifications, palaces, towns, temples, theatres, baths and aqueducts, the number of publications about Roman road and bridge construction is limited. There are works on individual aspects of this vast field, but they hardly go beyond general data on the time of construction, their location and characteristics. Until recently (O'Connor, Cambridge, 1993) there is hardly a systematic work on the subject in the English language. This is astonishing, since up till our time the Romans must be considered the foremost builders of roads and bridges in history.

Until recently, the study of engineering in antiquity has been neglected by historians. Politics were regarded the central theme of historical research (White, 1986). This derives from the fact, that archaeology itself only slowly emancipated from the grip of classical philology. Only since the middle of the 19th-century archaeology became a science in its own right and no more served philology. Between 1842 and 1845 the first chairs of archaeology were established in Göttingen, Berlin and Halle. Beyond historical and artistic considerations, the archaeologists soon faced the technical aspects of ancient buildings, which they had difficulty to appreciate and therefore tended to neglect.

Not before the middle of the 20th century begins the study of antique engineering. Up till then individual engineers studied particular inventions or achievements, including Roman bridges (H. Schneider, 1992).

Antiquity did not divide engineering into several branches, as we do today; antique engineers were all-rounders (White, 1986). In antiquity the art of building (*architectura*) encompassed building construction, mechanical and military engineering (*poliorcetica*). Thus *De Architectura* by Vitruvius includes chapters on machinery, clocks, catapults or even astronomy, besides the major part on building construction.

While technical progress was very slow during the Stone Age and even in the river civilizations of Egypt and Mesopotamia, the Greeks and Romans introduced important changes in construction technology. New implements and methods by application of newly discovered mechanical laws, allowed a much more effective use of human or animal power. However, the many competing city states of Greece were not able to engage much in public works such as roads or bridges.

It was Roman determination to conquer and develop, what they considered, inferior populations. The strategic mind of the Roman recognized that efficient communication over land and sea was the foundation of empire building. Their genius for organization brought with it effective administration, superior engineering and faultless execution of the works. It is in this order of importance that I shall organize this study, from the top down.

2. Building roads and bridges for an empire

Three elements are necessary for a road to be built: A commissioner to order it, engineers to plan and design it and contractors to build it.

2.1. Historical sources

Pliny the Elder (AD 23–79) prepared an encyclopedic *historia naturalis*, collected from Greek and Roman specialists in art, geography, medicine, zoology, botany and mineralogy.

Plinius Caecilius Secundus (the Younger), being *curator alvei Tiberis et riparum et cloacarum urbis* (administrator of the Tiber River, its embankments and of the city drains) had to deal with building problems. As governor of Bythnia he reported to Emperor Trajan many technical construction problems.

Pappos, a Greek mathematician from Alexandria prepared around AD 300 a compilation of mechanical details for lifts and cranes, the windlass and reel, pulleys, worm gears and gear shifts. Apollodorus of Damascus was »imperial architect« under Emperors Trajan and Hadrian (2nd c. AD). Prokopius mentions that he wrote a report about the bridge over the Danube near Turnu Severin at the Iron Gate in Romania, which he built in AD 104/105.

Prokopius of Kaisareia (Palestine) prepared around AD 500 to 560 by order of Emperor Justinian a book about buildings: *de aedificiis.* It includes the great bridge near Adapazari in northwest Turkey.

Also Anthemios of Tralles and Isidoros of Milet, who built the Haghia Sophia (532–537), have written technical reports, now lost. Most likely they new Heron's work *kamarika,* which must have been of great benefit for them.

This proves how the achievements of engineers were not only based upon apprenticeship and personal experience, but that there was a body of technical literature for reference.

The writings of Tacitus (AD 55–116), Sueton (AD 75–150) and Cassius Dio (AD 150–235) show how impressed they were by the outstanding engineering achievements. Cassius Dio gives as example the 700 m long road tunnel at Naples.

Plutarch writes in his biographies of great Greek and Roman personalities under Gaius Gracchus: »He was most diligent in road construction, considering pleasing design as much as technical quality. The roads stretch through the countryside in straight lines, low parts being filled, high parts cut, ravines and rivers bridged over and the embankments standardized, giving the whole a uniform and beautiful aspect. Each section was marked by milestones.«

Even in poetry we find an extraordinary source on road construction. The collection of poems by Publius Papinius Statius (*silvae* IV,3) from the 1st c. BC includes detailed technical descriptions. In *silvae* II, 2.5 he rhapsodizes in praise of Emperor Domitian: »Here nature yielded, defeated by the builder and docilely accommodated hitherto unknown service«. However, as far as bridges are concerned, it is astonishing how little impact they made on antique writers. In the year AD 371 the eminent scientist and poet Magnus Ausonius describes in his poem *Mosella* (Stuttgart, 2000) the beauty of the Mosel region, its towns, castles, palaces, villas and baths surrounded by vineyards – but he does not mention any of the many, beautiful bridges spanning the river. In modern times began the scientific study of antique technology, as a whole or as selected topics. Roman road and bridge construction did not receive much attention, one of the earliest source is Curt Merckel's *Die Ingenieurtechnik im Altertum* (the engineering technology in antiquity) from 1902.

2.2. Commissioning and maintenance

In Italy consuls were mostly responsible for road construction, while in the provinces praetors and proconsuls are involved. Wiseman (1970) lists the name and rank of 20 road builders in Republican times, consisting of 12 consuls, 6 censors and 2 praetors. According to Pekary (1968) censors were not eligible as public road builders, since they did not have the *ius publicandi,* which gave the right to expropriate private property.

Since Augustus the emperor was the supreme building authority. There was an imperial planning office handling large projects, to be approved by the emperor, who had an advisory committee of specialists.

The exchange of letters between Trajan and C. Pliny the Younger, governor of Bithynia, is a good example of the detailed involvement of the emperor (C. P. Secundus, *Letters to Emperor Trajan*). In letter 39 Pliny asks the emperor to send an expert to examine construction defects in a theatre in Nicaea (today Iznik in Turkey) and in baths under construction in Claudiopolis (today Bolu in Turkey). The emperor rejects the request, arguing: »You can't lack architects. There is no province, which hasn't experts and talented people (*peritos et ingeniosos homines*)«. In letter 41 he reports on the canal construction and asks for the provision of a leveling specialist and a hydraulic engineer.

The Roman emperors had a staff of technical advisors. On his extended voyages through the empire, Hadrian took a team of specialists (*fabri*), surveyors (*perpendiculatores*) and civil engineers along, to check on projects and facilitate their progress.

The roads were usually named after their commissioner, either a high official, a successful general or the emperor himself. Some of them considered it an honor to pay for »their« road or bridge, which could be rewarded by the right to add a triumphal arch, to mint commemorative coins, or to place statuary and memorial plates.

Many mile stones and bridges carry inscriptions such as *pontes et vias refecit, restituit.* On the *Pons Fabricius* in Rome, built in 62 BC to connect with the Tiber isle, are two inscriptions. The first reads: »Lucius Fabricius, son of Caius, being in charge of road (and bridge) construction, commissioned and accepted this bridge« [*L(ucius) Fabricius C(ai) f(ilius) cur(ator) viar(um) faciundum coeravit eidemque probaveit*]. The second inscription refers to repairs, which were necessary after the floods of 23 BC: »Marcus Lollius, son of Marcus, (and) Quintus Lepidus, son of Manius, received the works as consuls upon the Senate's decree.« *M(arcus) Lollius M(arci) f(ilius), Q(uintus) Lepidus M(ani) f(ilius) co(n)s(ules) ex s(enatus) c(onsulto) probaverun(t).*

In AD 20 Augustus was given the official authority over all road construction (*cura viarum*). With abundant funds he could order and build highways in the whole empire. In the provinces maintenance was in the hands of governors (*legati Augusti pro praetore*), in Italy it was the task of the *curators viarium,* a step up in the administrative hierarchy (*cursus honorum*).

Strabo relates at the turn of the millennium conditions on the Via Julia Augusta: »The city of Nîmes lies on this road, which leads from Spain to Italy; in summer this road is in good shape, but in winter and spring it is muddy and flooded by rivers. Some of them can be crossed by means of ferries, some over bridges of wood or stone.«

At the end of the Republic political trouble hindered road construction, but Augustus set out to repair infrastructure, especially the *via Flaminia* to the north: *»Consul septimum viam Flaminiam ab urbe Ariminum refeci pontesque …«* (For the seventh time, I, the consul have repaired the Via Flaminia and the bridges of Rimini) (Augustus, *Res gestae, Monum. Ancyr. 20*). However, soon after the poet Martial (IX, 57) of the 1st c. AD complains: *Nihil est tritius nec quae Flaminiam secant salebrae* (Nothing is more derelict than the potholes of the Via Flaminia)

2.3. The planning and design of roads and bridges

2.3.1. Generalities

The Romans had a particular genius in determining the routing of roads. Over-all they followed strategic considerations, but locally they chose the best position of a road, be it in the valley, along a hillside or on a plateau. They avoided snowed-in northern slopes, flooding zones and swamps. On the other hand, they did not avoid major obstacles, if they felt it worthwhile. The Via Appia features 10 m high buttressing walls around Ariccia; near Terracina a 40 m high rock projection was removed; in the Aosta valley there are cuttings, which retained milestones and a triumphal arch out of solid material.

They also straightened out existing republican roads, such as at the Ponte di Nona on the Via Praestina (RE XXI, Sp. 2432). The low single arch bridge was replaced by a much higher bridge of seven arches.

There are a few tunnels for carriage ways, such as the 700 m tunnel *crypta Neapolitana* near Naples. Seneca describes in a letter the problems of the »rush hour«: »Nothing is darker than this torch light, we can't distinguish the shadows and if there would be more light, it would be extinguished by the dust« (letter to Lucilius LVII).

At Forlo near Pesara the Via Flaminia has a 40 m tunnel cut out of lime stone, which is still in use.

In the mountains considerable grades had to be accommodated. The Romans preferred to cover great differences in levels by steep but short ramps, instead of maintaining a winding road with uniform slope. At the Maloja Pass to Switzerland ropes and winches were used to pull carriages up. When travelling to Gaul, the Romans preferred the straightforward but 1850 m high Mont Genève pass, to the ever winding road along the Mediterranean.

2.3.2. Antique literary sources

Existing literary sources yield practically nothing about Roman bridge building (White, 1986). Herodot (I 186), Diodor (II 8, 2–3) and Curtius Rufus (V 1, 4) report the construction of a bridge over the Euphrates in Babylonia at about 600 BC. War chroniclers describe the construction of temporary military bridges, generally built with timber or over pontoons. Best known are Caesar's bridges over the Rhine in the first c. BC, or a pontoon bridge over the Bosporus, built in 513 BC by Mandrokles of Samos for the Persian King Dareios (Herodot IV. 87–88).

In contrast there are numerous records about the history of mechanical technology, hydraulics, surveying, water distribution, agro-technology, tunneling, shipbuilding, military technology and general construction methods.

2.3.3. Formation of architects and engineers

Our best Roman source on building matters are the *Ten Books on Architecture* (*De architectura libri decem*) by Vitruvius Pollio. He was born in 89 BC and received profound training as an architect – in today's sense equal to an engineer's formation. He served as siege specialist in the armies of Caesar and Augustus. Vitruvius was a man of practice. During his retirement he wrote the classical work on Roman construction. In ten sections he discusses the architect's formation, properties of materials, the design of temples, markets, basilicas, theatres and private dwellings, as well as the preparation of lime and screeds, the decoration of buildings, and even hydraulic and mechanical engineering, surveying, siege technology and astrology.

Vitruvius tried to depict Greek and Italic-Etruscan architecture as variants of one whole. He praises the buildings of the late Republic and criticizes contemporary innovations such as vaulting, brick or multi-floor construction.

Unfortunately he was too early to cover the use of concrete (*opus caementitium*), the revolutionary and dominating construction technique of imperial times. Roman concrete attained strengths of today's product. Neither does he treat large span construction by means of vaulting with voussoirs, thus ignoring a technique basic for bridge building. Briefly he mentions the use of buttressing arches or arches as substructure for buildings. It is difficult to tell, why he does not at all mention the construction of bridges, although there were already outstanding examples during his time. Nonetheless his treatise is of great value for this study, as it covers materials and the preparation of mortars, foundations, hydraulic works and methods of drainage.

However, we must avoid seeing his compendium as a comprehensive work of all antique building knowledge – as architects maintained during the Renaissance. After all, the most impressive Roman structures were created after his death. Not until the 15th century his writings were rediscovered in the Swiss monastery of St.Gall.

Vitruvius lists the many requirements of an architect: The architect's knowledge derives from practical work (*fabrica*) and scientific study (*ratiocinatio*). He must be accomplished in both ways. Neither talent without schooling, nor schooling without talent can produce a master. He must be a skilful writer and draftsman, should have studied geometry, history and philosophy, like music, know some medicine, law and astronomy, be familiar with the various seasons.

In chapter I/11 he writes: »Consequently, since this study is so vast in extent, embellished and enriched as it is with many different kinds of learning, I think that men have no right to profess themselves architects hastily, without having climbed from boyhood the steps of these studies and thus, nursed by the knowledge of many arts and sciences, having reached the heights of the holy ground of architecture.«

In antiquity the term architect had a much wider meaning than today. It derives from the Greek *archi* = chief and *tekton* = builder and meant chief- or master builder. At this time it included engineering in general, specifically civil engineering. The term engineer derives from *ingenium*, something ingenious, clever. He works on the basis of applied sciences, to make something work, while the architect's work should satisfy the requirements for utility, solidity and beauty. The exalted image of the architect's profession not always matched Roman reality. For instance, an in-

1. Sepulchral stele of *mensor aedificiorum* Titus Statilius Aper, Capitoline Museum, Rome.

2. Top: angle gauge from the Avignon Museum (France) to mark 45° angles, bottom: gauge from the Louvre to determine angles, vertical and horizontal planes.
3. Bronze compass and plumb bob from Swabia (AD 1st to 3rd c.)

scription in Lambèse (Algeria) dated AD 152 tells us: Returning after a lengthy absence to a tunnel construction site, the military engineer of the local *Legio III-Augusta* discovered, that in spite of exact instructions and commandments, the contractor made so many mistakes, that the tunnel parts driven from either side did not meet in the middle. If he had not come back in time, the client would have had two tunnels instead of one.

2.3.4. Preparing plans

How did the Roman architect/engineer document and communicate his plans? Vitruvius tells us (I, 1, 4) that ground plans, elevations and perspectives, including dimensioning, were part of the building trade. Large buildings were often defined by a model. Unfortunately, the drawings which had been part of his *de architectura* have been lost.

The Roman architects used papyrus, parchment or wax tablets to draw on, but none of this survived. The tomb of Titus Statilius Aper, *mensor aedificiorum*, shows the deceased with a parchment roll and a bundle of wax tablets, plus a quiver with styluses. (Grewe, 1985)

Design drawings were made on parchment, while the wax tablets served for sketches or notes. Usually, working dimensions were laid out on top of foundations, floor slabs, screeds or timber platforms at full scale, to guide the workmen.

The exact preparation of stones, such as for arches or vaulting, required detailed plans and measurements. Especially the wedge like shapes for vaulting needed to be defined by jigs and templates. Usually, only the bearing surface was finished. Whole arches were laid out like this, to be assembled on site. Exposed surfaces and decorations were finished on the erected structure.

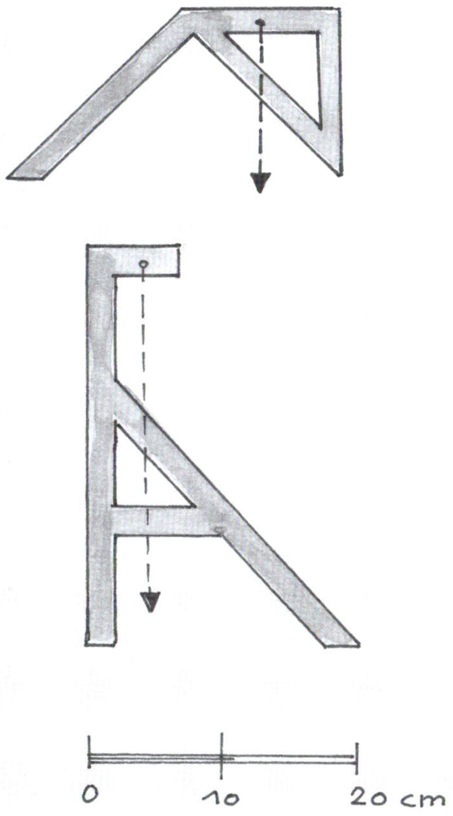

2.3.5. Surveying

Already Gaius Julius Hyginus (around 64 BC to AD 17) wrote about astronomy and the art of surveying. The most famous source is Sextus Julius Frontinus (around AD 30–104). As *curator aquarum* he was in charge of Rome's water supply, which he described in *de aquis urbis Romae.* He also wrote about the art of field surveys (*corpus agrimensorum*), about warfare and stratagems (*stratagema*).

2.3.5.1. Measuring units

The basic Roman unit of measurement for building was the foot (*pes*). Many subdivisions were in use. J. P. Adams (1984) gives the following details:

digitus	1/16	foot =	1.848 cm
palmus	1/4	foot =	7.392 cm
pes	1	foot =	29.570 cm
palmipes	1 1/4	foot =	36.960 cm
cubitus	1 1/2	foot =	44.355 cm
gradus	2 1/2	foot =	73.925 cm
passus	5	foot =	1.478 m
decempeda	10	foot =	2.957 m
mille passus	5000	foot =	1478.500 m

Further measures in use were the *pertica*, the third of one *decempeda* = 98.566 cm. There were calibration boards, such as one in Thibilis (Algeria), indicating the Roman foot and other yardsticks. (Grewe, 1982)

2.3.5.2. Layout, measuring and leveling implements

Set squares of wood or bronze were employed to produce a right angle – for the horizontal layout of right angles the rule of Pythagoras was used, with a rope, knotted at 3, 4 and 5 foot. Water levels allowed the control of horizontality.

The Roman surveyors were variably called *agrimensores, metatores, mensores* or *libratores* (levelers). For linear measurements they used a pole or waxed ropes with markings. For leveling and determination of ground angles they used the chorobates, the groma and the dioptra. Grewe remarks, that the method of leveling hardly changed since antiquity. It all depends upon accurate horizontal sightings, for which served the chorobates.

Next to the common water level (*libra aquaria*) the chorobate was the most important tool for establishing alignments and levels. It consisted of a wooden frame with a horizontal 20 ft straight edge, fitted with a water level. At the ends were plumb bobs to check verticality. A groove served as sight line. It had a notch and bead sight to establish distant levels. Vitruvius writes on this matter: »First comes the method of taking the level. Leveling is done either with dioptrae, or with water levels, or with chorobates. It is done with greater accuracy by means of chorobates, because dioptrae and levels are deceptive. The chorobate is a straightedge about twenty feet long. At the extremities it has legs, made exactly alike and jointed on perpendicularly to the extremities of the straightedge and also to crosspieces fastened by tenons, connecting the straightedge and the legs. These crosspieces have vertical lines drawn upon them and there are plumb lines hanging from the straight edge over each of the lines. When the straightedge is in position and the plumb lines strike both the lines alike and at the same time, they show that the instrument stands level.

... But if the wind interposes and constant motion prevents any definite indication by lines, then have a groove on the upper side, five feet long, one digit wide and a digit and a half deep,

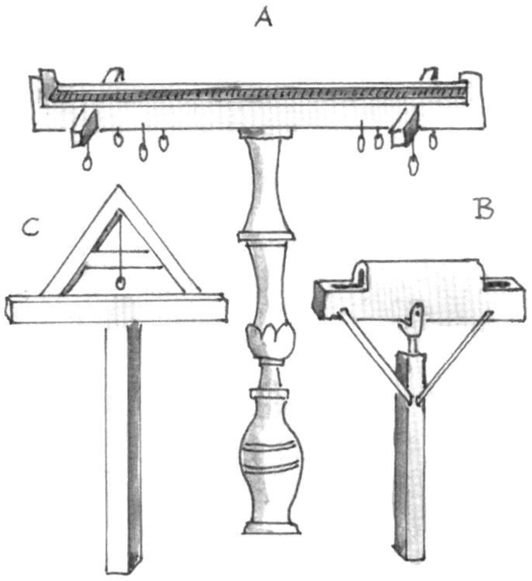

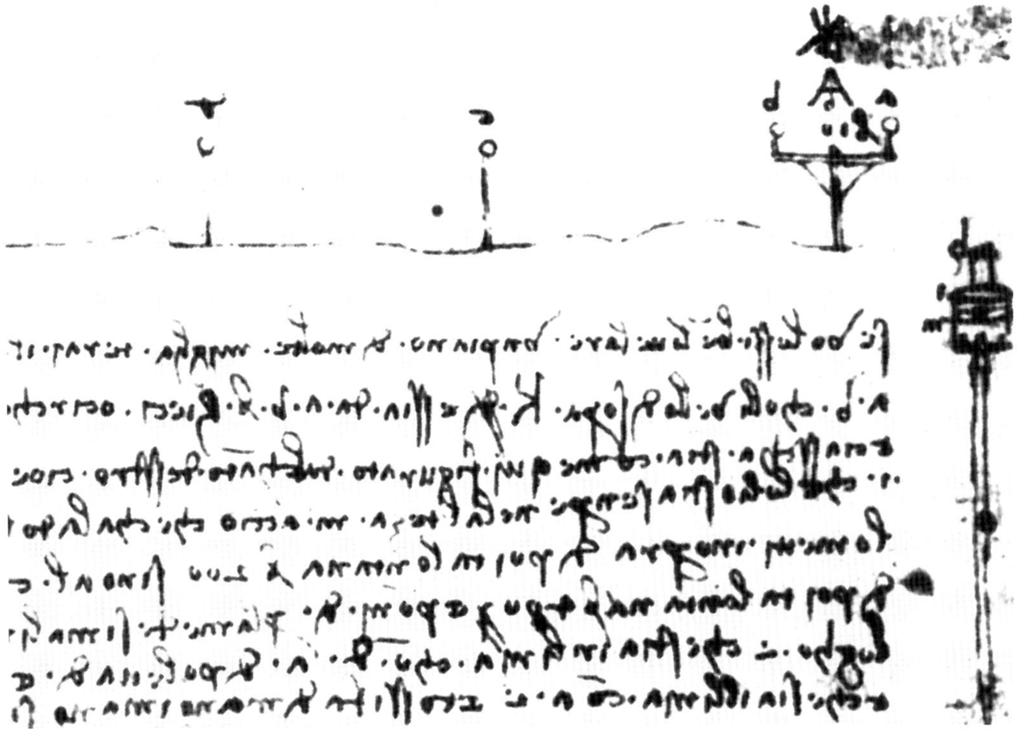

4. Leveling tools after Fra Giovanni Jocundus, Dominican friar and architect (Italy, 1511). (A: chorobate, B: libra aquaria (water level), C: dioptra.

5. Sketch of a chorobate by Leonardo da Vinci.

6. Reconstruction of a chorobate by Neuburger (1929).
7. Reconstruction of a chorobate by Adam (1984). Wasserkanal = water channel, Richtscheit = straightedge, Hängelot = hanging perpendicular, Markierung = mark.
8. Reconstruction of leveling tools by Cesare Cesarino (Como, 1521).
9. Taking levels by two-way sighting.
10, 11. Reconstruction of a chorobates by Dilke (1971). Meßlatte = level rod, Visier = sight, Wasserrinne = water channel, Strichmarke = mark, Hilfslot = auxiliary perpendicular.

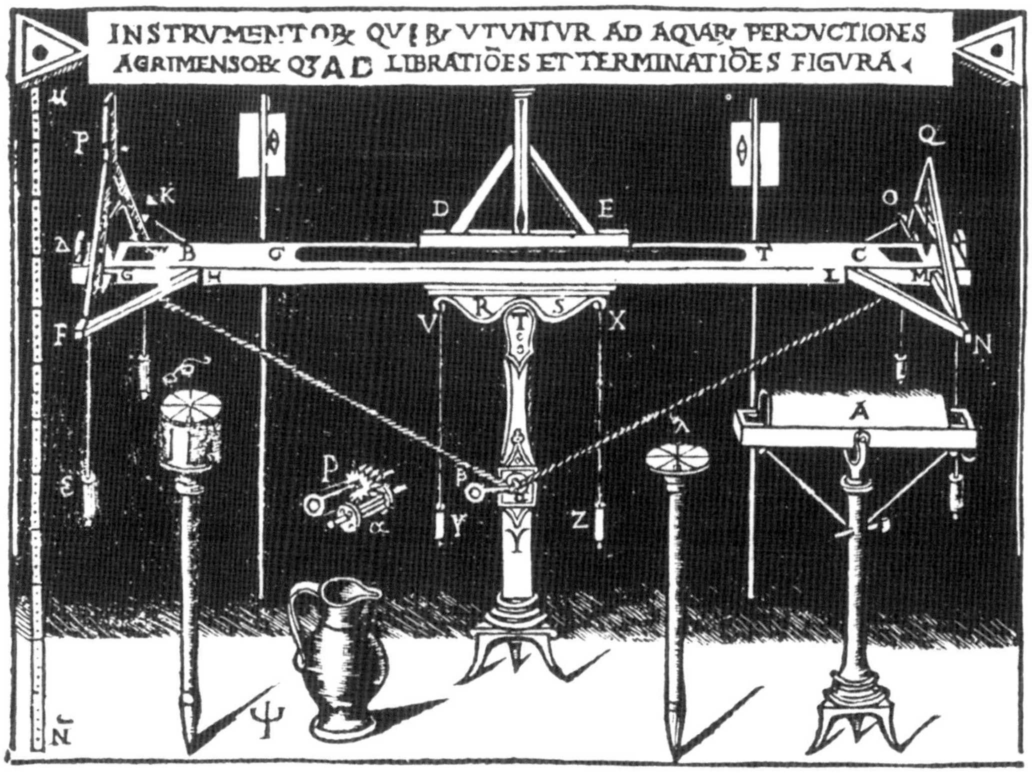

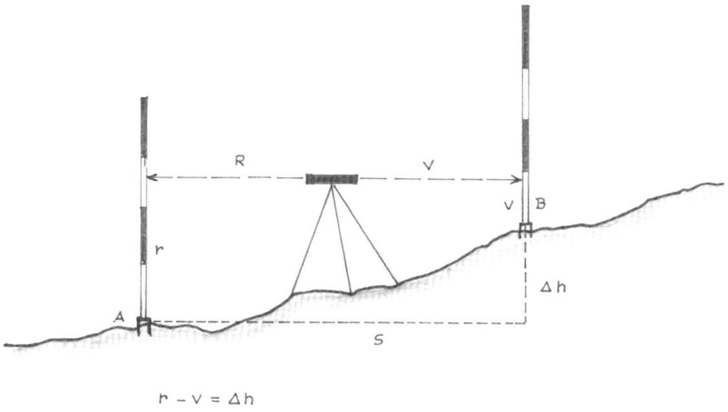

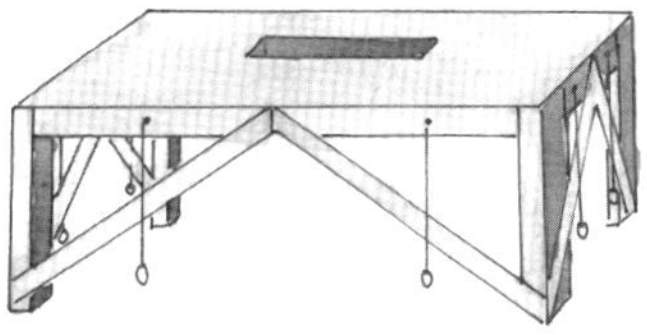

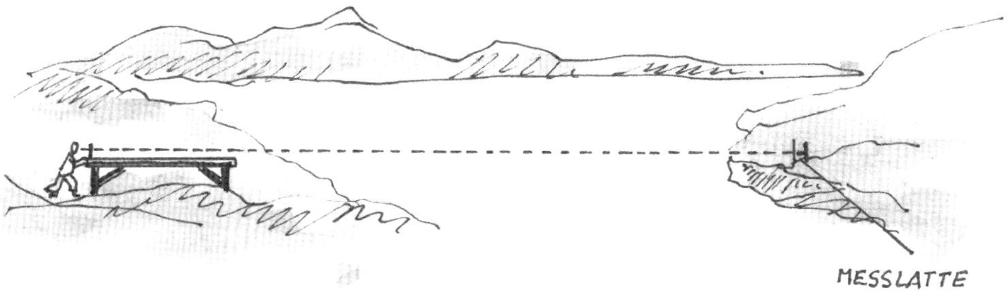

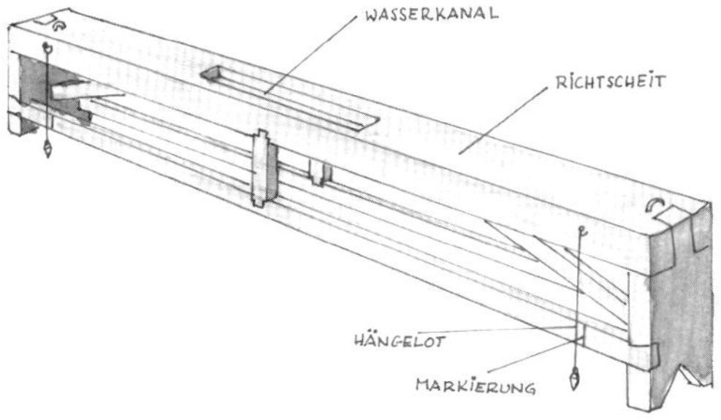

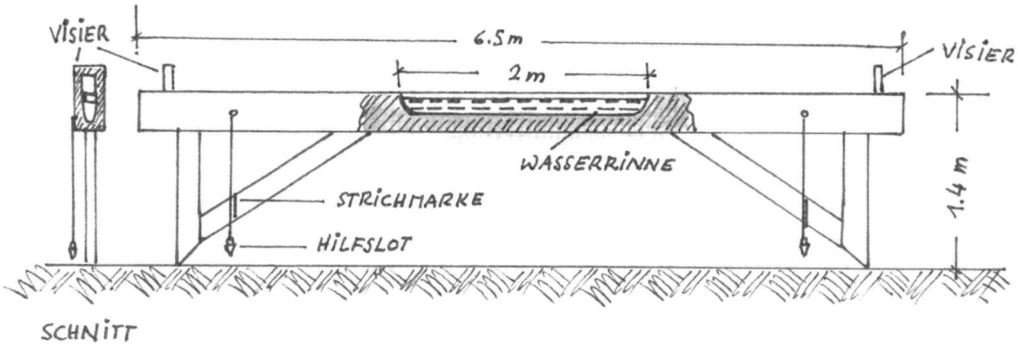

and pour water into it. If the water comes up uniformly to the rims of the groove, it will be known that the instrument is level. When the level is thus found by means of chorobates, the amount of fall will also be known.«

They were especially used for the construction of water mains and aqueducts. For instance, the 50 km water supply of Nîmes across the Pont du Gard had a continuous gradient of 34 cm per km, an excellent achievement. Since the drawings of Vitruvius are lost, there have been several attempts to reconstruct the chorobates.

The groma served for measuring angles. On a stand or tripod a revolving cross of coordinates was fitted with plumb lines. Sightings across the plumb lines allowed the marking of angles. The marking of a right angle was done by sightings across the two sets of plumb lines.

The diopter (*dioptra*) was similar to our transit. Already Heron of Alexandria (1st c. AD) described this precision instrument, which allows to establish horizontal and vertical angles. An integrated water level of communicating vessels assured the horizontal rotation of a sighting device.

An example of the perfect execution of leveling is the water supply of Samos at the time of Polycrates (535–522 BC). An 825 m long tunnel had to be cut through a mountain. To speed up the work one proceeded from both ends. Meeting in the middle, the two teams were only 1 m off vertically.

12. Reconstruction of a groma found in Pompeii.
13. Model of a groma in the National Museum of Naples (White 1984).
14. Groma shown on the sepulchral stele of the agrimentor Nicostramus in Pompeii (Adam 1984).

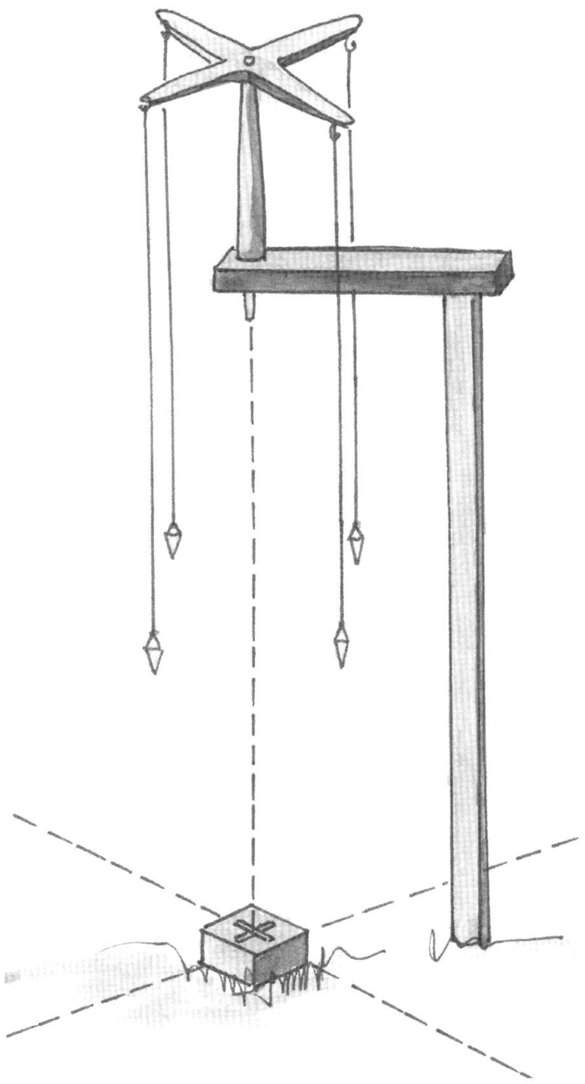

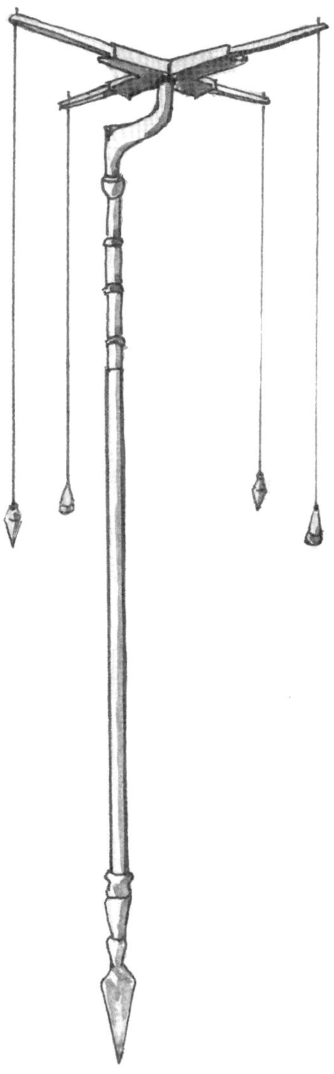

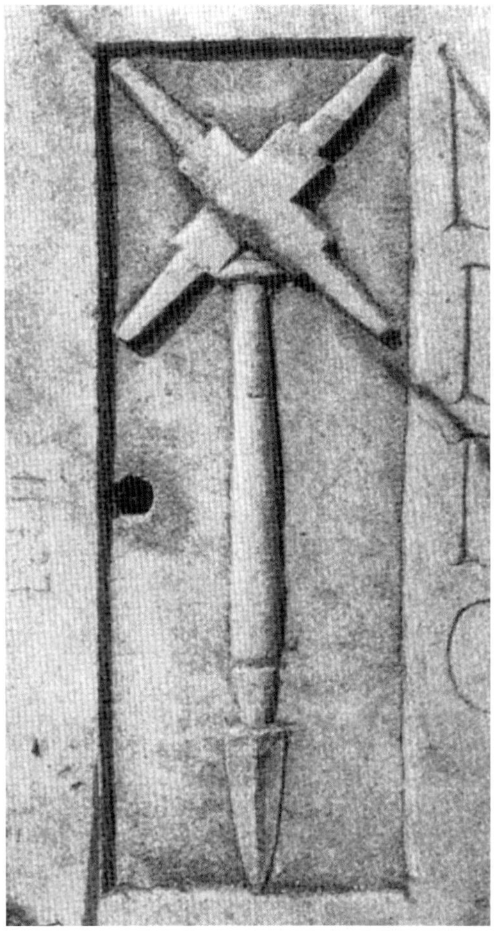

15. Model of a first century groma at the Saalburg, Germany (Kretschmer, 1983).
16. Roman surveyor with a groma (Kretschmer, 1983).
17. Tombstone of the mensor Lucius Aebutius Faustus from Ivrea (Italy) (Greve, 1985).
18. Reconstruction of a *dioptra* according to the description by Hero of Alexandria (White, 1986).
19. Auxiliary attachment to Hero's diopter on the basis of communicating vessels. Two vertical tubes connected by an internal channel indicated horizontality (Neuburger, 1987).

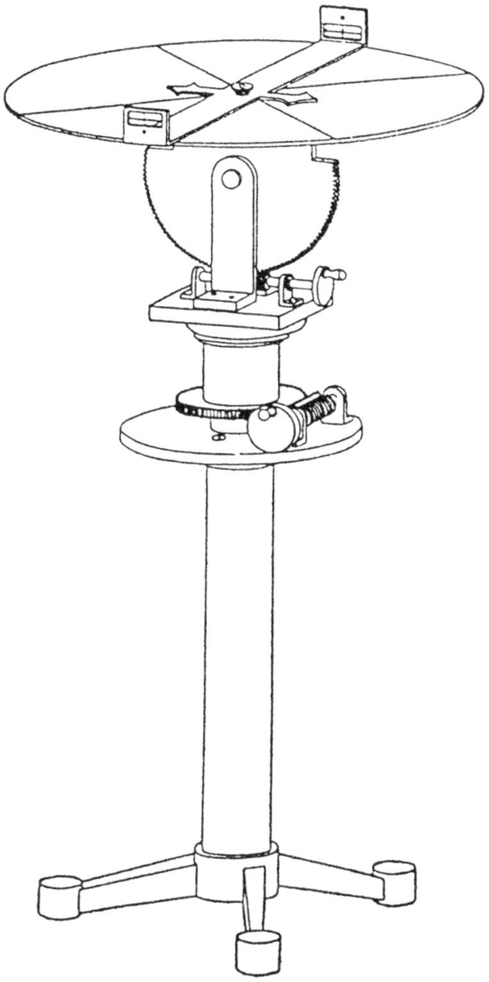

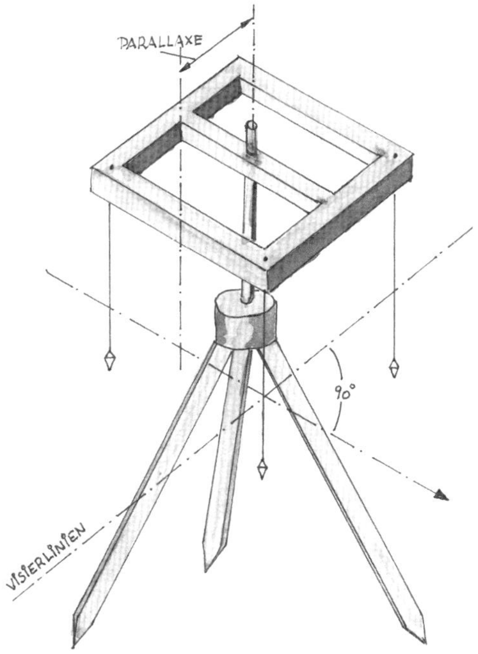

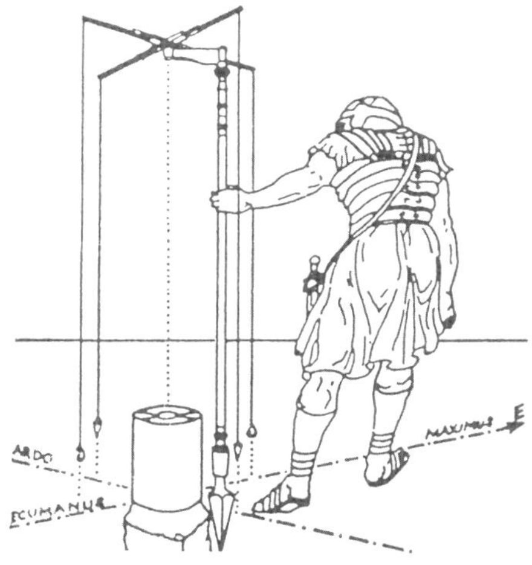

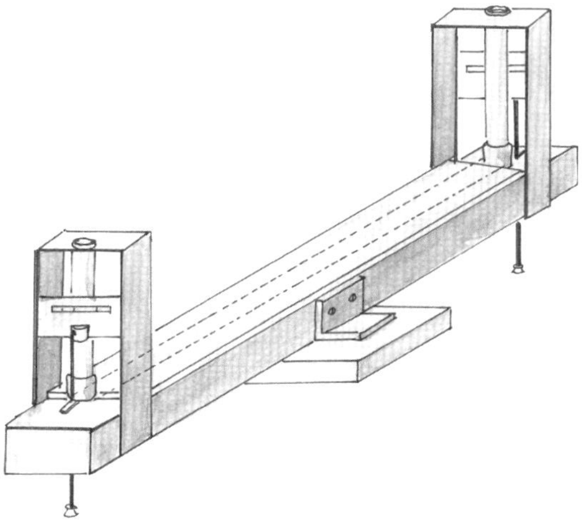

2.3.6. Materials and their use

2.3.6.1. Building materials and their procurement

The fabric of roads and bridges consisted mainly of stone, concrete (*opus caementitium*) and burnt brick. In addition bronze, lead and wood were used as connectors and temporary shuttering. Quarries give us clues about the extraction and quality of materials, transport problems, the number of workers and the duration of construction. Another source are the remains in place, or reused parts of masonry in other buildings. Modern material testing methods also offer new knowledge. Later we shall return again to materials.

While for small, private projects the material was furnished by the local builder, the procurement of materials for large, public projects was very early on managed by the government. Since imperial times the state was the dominant client in construction and most quarries – especially for marble – as well as brickyards and mines became government property.

2.3.6.1.1. Stone

The most important building material was stone. Vitruvius discusses at length the necessary properties of stone for building (II, chap. 7), revealing the ancients' notion of the components of materials:

»1. Next comes the consideration of stone quarries from which dimension stone and supplies of rubble to be used in building are taken and brought together. The stone in quarries is found to be of different and unlike qualities. In some it is soft: for example, in the environs of the city at the quarries of Grotta Rossa, Palla, Foidanae, and of the Alban hills; in others, it is medium, as at Tivoli, at Amiternum, or Mt. Soracte, and in quarries of this sort; in still others it is hard, as in lava quarries. There are also numerous other kinds: for instance in Campania, red and black tufas, in Umbria, Picentum and Venetia, white tufa which can be cut with a toothed saw, like wood.

2. All these soft kinds have the advantage that they can be easily worked as soon as they have been taken from the quarries. Under cover they play their part well; but in open and exposed situations the frost and rime make them crumble, and they go to pieces. On the seacoast, too, the salt eats away and dissolves them, nor can they stand great heat either. But travertine and all stone of that class can stand injury whether from a heavy load laid upon it or from the weather; exposure to fire, however, it cannot bear, but splits and cracks into pieces at once. This is because in its natural composition there is but little moisture and not much of the earthy and watery elements, but when fire, expelling the air from it by the operation and force of heat, pene-

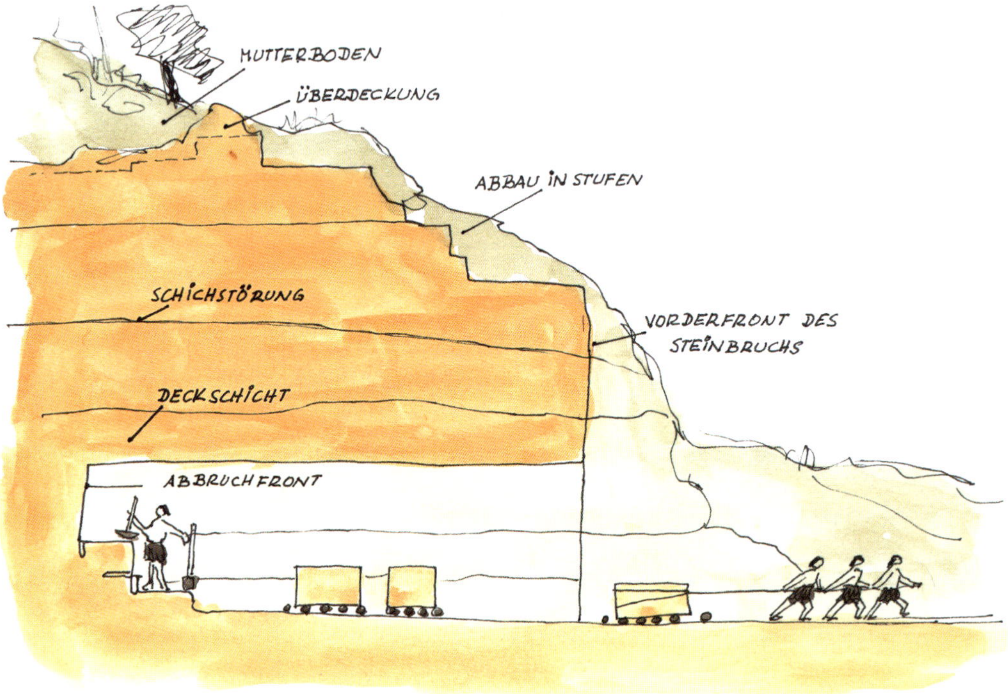

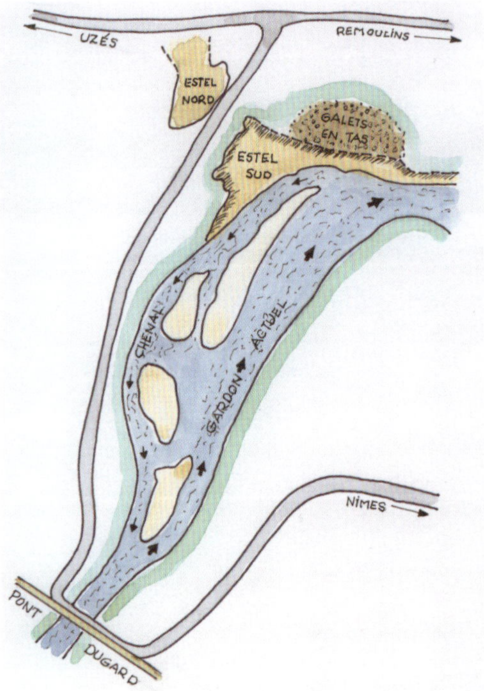

trates into its inmost parts and occupies the empty spaces of the fissures, there comes a great glow and the stone is made to burn as fiercely as do the particles of fire itself.

3. There are also several quarries called Anician in the territory of Tarquinii, the stone being of the color of peperino. The principal workshops lie around the lake of Bolsena and in the prefecture of Statonia. This stone has innumerable good qualities. Neither the season of frost nor exposure to fire can harm it, but it remains solid and lasts to a great age, because there is only little air and fire in its natural composition, a moderate amount of moisture and a great deal of the earthy. Hence its structure is of close texture and solid and cannot be injured by the weather or by the force of fire.

4. This may be best seen from monuments in the neighborhood of the town of Ferento which are made of stone from these quarries. Among them are large statues exceedingly well made, images of smaller size and flowers and acanthus leaves gracefully carved. Old as these are, they look fresh as if they were only just finished. Bronze workers, also, make moulds for the casting of bronze out of stone from these quarries and find it very useful in bronze-founding. If the quarries were only near Rome, all our buildings might well be constructed from the products of these workshops.

5. But since, on account of the proximity of the stone quarries of Grotta Rossa, Palla and the others that are nearest to the city, necessity drives us to make use of their products, we must proceed as follows, if we wish our work to be finished without flaws. Let the stone be taken from the quarry two years before building is to begin, and not in the winter but in summer. Then let it lie exposed in an open place. Such stone as has been damaged by the two years of exposure should be used for foundations. The rest, which remains unhurt, has passed the test of nature and will endure in those parts of the building which are above ground. This precaution should be observed, not only with dimension stone, but also with rubble which is to be used in walls.«

Around Rome the following kinds of stone were used (Lugli, 1957):

Peperino (*lapis Albanus*), a gray tuff, containing fragments of lava and lime stone. It was used for the Via Appia's bridge across the Ariccia valley.

Lapis Gabinus, a grey-green tuff from Gabi, 19 km southeast of Rome and similar to peperino. It was used in the bridges Aemilius, Milvius and Fabricius, as well as the *Cloaka Maxima.* Other tufa were also named after their origin, such as Tufo d'Aniene or *Lapis Pallens*.

Travertine (*lapis Tiburtius*) is a very hard form of lime stone deposited by hot springs. It is cream colored with an interesting structure and ideal for stone cladding.

Marble was hardly used for construction because of its high value. Sometimes keystones of marble accentuated their importance.

A crucial element was the transportation of stone and any suitable material close to the site was used. This includes igneous rocks such as granite, syenite, diorite or gabbro. Their high compressive strength of 1000 to 3000 kp/cm^2 made them ideal for arch construction.

Volcanic rocks (700–4000 kp/cm^2) such as porphyry, trachyte or basalt were very abrasion and water resistant and therefore mostly used for paving.

Sedimental rocks such as sand- and limestone with a strength of 200 to 1800 kp/cm^2 have been used wherever available. According to O'Connor (1993) 90 % of the bridges in Italy were built with travertine, tuff or limestone.

Porous volcanic tuffs of comparable strength but as little weight as 800 kp/cm^2 are particularly suitable for vaulting. They were often used for the upper parts of vaulting to reduce weight and thrust.

The strength of a stone is not necessarily an indication of its durability. It depends upon its crystalline structure and porosity. Vitruvius knew that some stone weathers very quickly at the contact with air after its extraction from the quarry.

2.3.6.2. Quarrying and transportation

Accessibility of rock from the ground is the key for economical quarrying. After all growth and topsoil is cleared, weathered material must also be removed until solid rock is reached. Excavation then proceeds in steps from the top down, in accordance with the desired height of blocks. Often it is preferable to tunnel horizontally into the face of the rock, along the seams of existing strata, or even dig down vertically and pull the stones up.

20. Example of subterranean horizontal extraction of stone near El Haouaria (Tunesia), called the *Rhar-el-Kebir*, large grottoes. The porous yellow limestone was used for the construction of Carthage. The underground quarries are still accessible, they are about 15 m wide and 25 m deep. (Mutterboden = topsoil, Überdeckung = cover, Abbau in Stufen = step-by-step quarrying, Vordergrund des Steinbruchs = face of quarry, Schichtstörung = seam of stratum, Deckschicht = ceiling layer, Abbruchfront = rear face of quarry.)
21. Example of vertical quarrying from El Haouaria from a model in the Antiquarium in Carthage (Rakop, 1997).
22. Proximity of a quarry was of great value. For the Pont du Gard very good stone was available 700 m downstream, where the material could be directly placed on rafts and brought to the site. Map by Bessac et al. (2002).
23. Views of a modern travertine quarry at Ascoli Piceno (Central Italy).

In the absence of natural stratification, horizontal and vertical grooves were cut, or series of holes drilled, according to the size of a block, and the block split off with iron wedges driven in. Sometimes dry wedges of wood were inserted between iron plates. By watering, the swelling force of the wood split off the block.
Sawing stone was a kind of grinding procedure, moving a metal blade back and forth along a groove, with very hard, watered sand in between. The ancient Egyptians accomplished astonishing things with such a primitive method. Ausonius (about AD 370) describes in his poem »Mosella« some flour mills along a tributary of the Mosel River, which drove marble cutting saws. And Gregor of Nysa (4th c. AD) claims in a description of quarries in Cappadocia that stone is being cut with iron and water.

How to change the rotary movement of a mill into a linear, horizontal action was already discussed by Heron of Alexandria in the 1st c. AD and should have been known to the Romans.

For bridge construction the weight of individual blocks hardly exceeded seven to eight tons. Larger blocks required for architectural parts, such as column shafts, lintels, cornices had to be excavated all around and finally split off along the bottom. The biggest blocks quarried in this fashion are found in Baalbek (Roman Syria now Lebanon). They form the famous Trilithon, three blocks of stone as the middle layer of the podium enclosure for the Temple of Jupiter. Each block measures 20 by 4.6 by 4 m and weighs close to 800 tons. They are perfectly placed on top of the blocks below, without any mortar joint. Nobody has come up with a convincing explanation of how the blocks were put into position. Another block has been readied in the quarry, weighing close to 1000 tons.

Ideally, the building site was lower than the quarry, offering the opportunity to bring the blocks down on ramps, either on rollers or sleighs. Otherwise special roads were built to allow transportation on carts or carriages. There were *clabularia,* heavy oxen carts taking loads up to 500 kg, lighter four-wheelers called *rada* pulled by mules (300 kg loads), *vereda* taking 100 kg and the two-wheeled *birota,* mainly for passengers.

Very big pieces were encased in wheel shaped timber constructions and rolled along, pulled by oxen. Vitruvius (X, 2, 11–13) writes of »towing machines« for columns and architraves.

24, 26. Cutting of grooves with a pick, in preparation for wedging off.

25, 27. Placing wedges by hand or with a jackhammer.

28. Roman quarry in the Odenwald, Germany. It was used until the 4th c. AD (Kretschmer, 1983).

29. Reconstruction of transport encasements as used by Chersiphron in 550 BC for the construction of the Temple of Artemis in Ephesos (Newton, 1791).
30. Transport of a stone block on a sleigh, secured by ropes.
31. Rolling device with the load as axle, used by Metagenes, the son of Chersiphron, to transport architraves and columns of the temple. Diameter of the wheels 4 m (White, 1984).

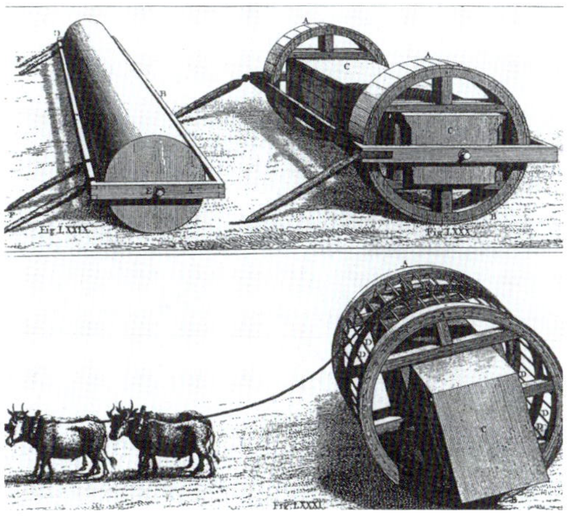

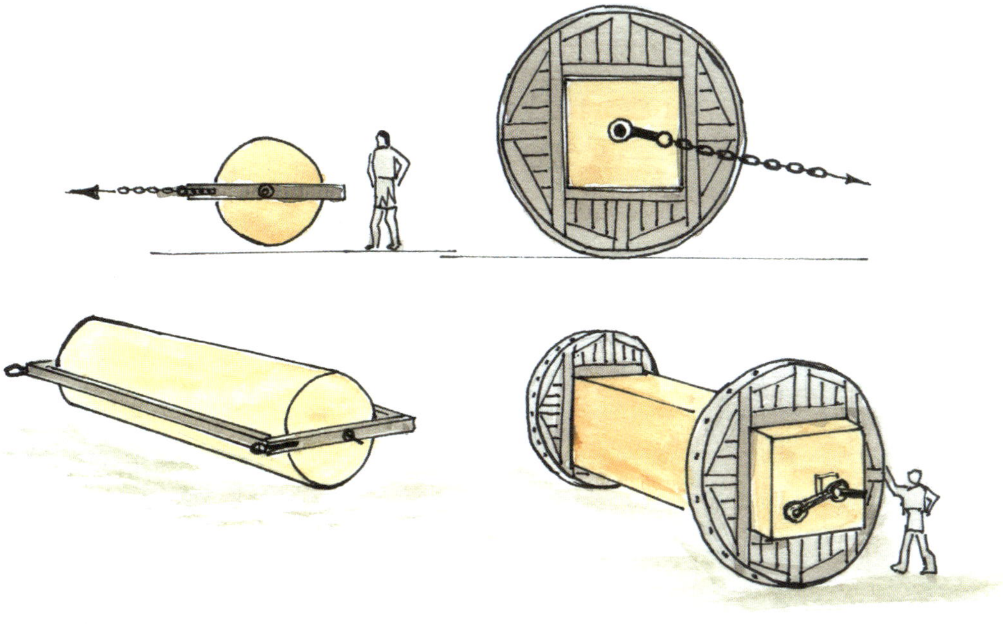

2.3.6.3. Masonry work

Cutting and dressing of stones was done on site. Ashlar masonry requires accurately squared blocks. To simplify fitting, only 3 cm wide edge strips were exactly finished and the inner surface slightly hollowed out (*anathyrosis*). The exposed face of the stone was left rough, for later finishing on the whole structure. Holes for cramps were cut in advance.

32. Prepared ashlar block (Anschlußfuge = top face to receive mortar, Stoßfuge = lateral joint with dressed edges and recessed inner face, ziselierte Umrandung = dressed frame and projecting bossed front). Note the lewis hole top center and the cut-outs on the sides for the attachment of a scissor or crowbars.
33. The stone mason's tools have not changed till today. They consist of maul hammer (1), pick (2), adze (3), hammer (4), pointed (5) scarifier (6), flat chisel (7), gouge (8) and angle gauge (9), made of iron or bronze.
34. Roman stone working tools.

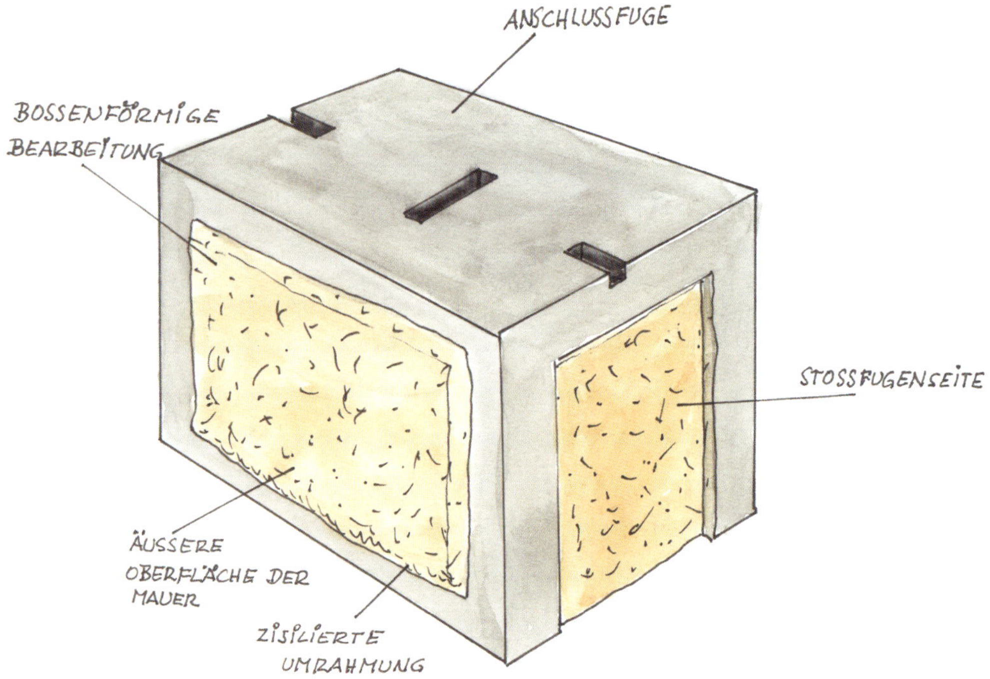

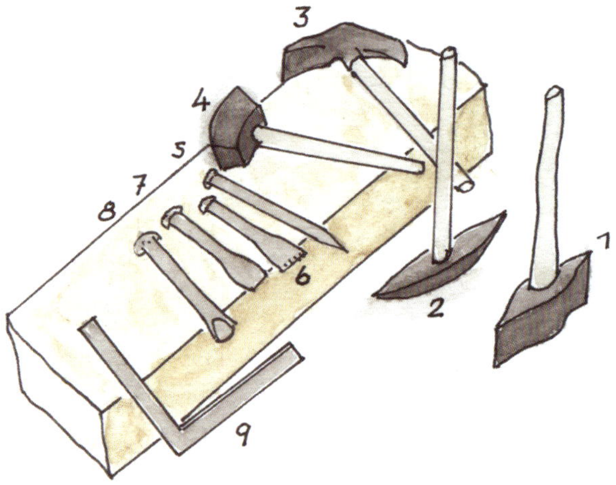

35. Opus quadratum isodomum: uniform masonry courses, runners only, alternate joints line up (mainly in Greece).
36. Opus quadratum isodomum: runners and tie stones alternate (mainly in Greece).
37. Opus quadratum pseudoisodomum: stone courses of variable height.
38. Opus quadratum with random length stones (mainly Etruscan).
39. Opus quadratum with alternating runner and binder courses (two faced work).
40. Opus quadratum as before, but single faced work.

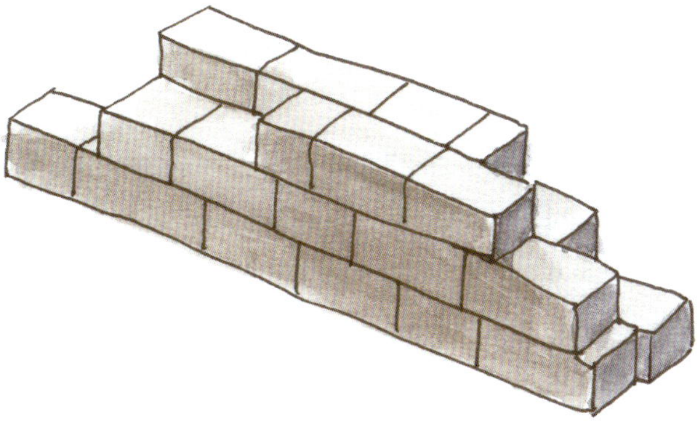

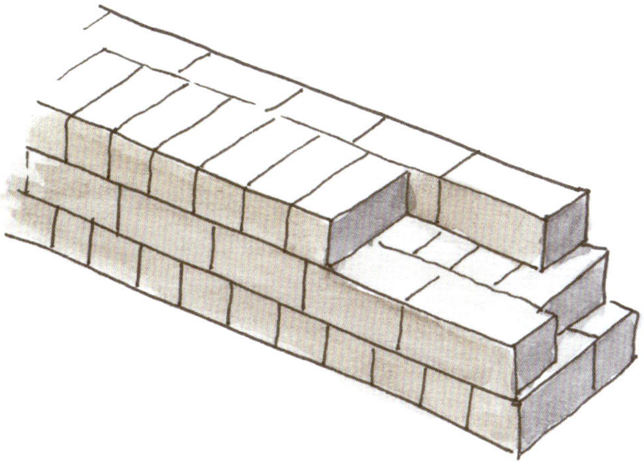

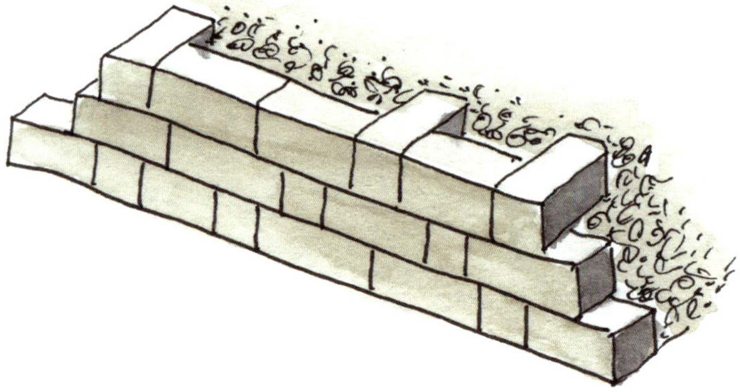

41. Stone facing for rubble core masonry (stone acts as lost formwork).
42. Same in heavier version.

In bridge construction *opus quadratum* = ashlar masonry was preferred. Squared blocks of up to 2.5 m length, 1 m height and up to 1.5 m thickness were placed without mortar. In early times cyclopean masonry consisted of irregular, quarry-shaped stones, the gaps filled in with smaller stones. For polygonal masonry the irregular stones were trimmed to fit adjoining blocks.

Roman concrete (*opus caementitium*) is one of the most important »inventions« of the Romans. The Latin word *opus* means an accomplished piece of work, *caementitium* derives from broken stone or rubble. Together it means »a work of rubble«. It was based upon the Roman discovery of hydraulic cement, which was found only in volcanic regions. With the fall of the Roman Empire the knowledge was lost. Cement was rediscovered as a side product of iron smelting in industrializing England, thus the brand Portland cement.

To quote Vitruvius again (II, 6): »There is also a kind of powder which from natural causes produces astonishing results. It is found in the neighborhoods of Baiae and in the country belonging to the towns round about Mt. Vesuvius. This substance, when mixed with lime and rubble, not only lends strength to buildings of other kinds, but even when piers of it are constructed in the sea, they set hard under water. The reason for this seems to be that the soil on the slopes of the mountains in these neighborhoods is hot and full of hot springs. This would not be so unless the mountains had beneath them huge fires of burning sulphur or alum or asphalt. So the fire and the heat of the flames, coming up hot from far within through the fissures, make the soil there light, and the tufa found there is spongy and free from moisture. Hence, when the three substances, all formed on a similar principle by the force of fire, are mixed together, the water suddenly taken in makes them cohere and the moisture quickly hardens them so that they set into a mass which neither the waves nor the force of the water can dissolve.«

Opus caementitium required formwork to support it in its plastic state. It could consist of wooden shuttering, which received on the inside a layer of small pyramid shaped stones, which bonded with the concrete and gave a finished surface when the shuttering was removed (*opus reticulatum*). Or the facing was of self-supporting masonry of stone or brick (*opus mixtum*).

2.3.7. Technology

In Roman times the use of construction machinery was very limited, as it was indeed till the 19th century. Work was done by hand, with hammer, chisel, adze, hatchet, pickaxe, drill, file, crowbar and spade.

In bridge building the employment of machines (*machinae*) and tools like pumps, rams, cranes and pulleys brought significant advantages, leading to the construction of bridges of hitherto unknown size. The increasing expertise of engineers and the high level of workmanship by the artisans allowed ever more daring solutions with far flung vaulting.

2.3.7.1. Implements, tools and machinery

This concerns items made of bronze, copper, wood or stone. The reconstruction of machines, cranes and tools for exhibition purposes gives us important insights into the practice of Roman builders. They are based on ancient written sources, such as by Aristoteles (383–322 BC) who wrote about *mechanica.*

Archemides (287–212 BC) discovered the laws of leverage and the center of gravity, he explained the effect of pulleys, of the screw and the inclined plane, all subjects of interest for bridge builders. He even determined the distance and size of celestial objects. Still existing are his works on spheres, cylinders and floating objects. His biggest contributions are in calculus and the laws of equilibrium.

Philon of Byzantium (2nd c. BC) authored the work *mechanike syntaxis.* Heron of Alexandria (1st c. AD) wrote about catapults (*belopoika*), air pressure (*pneumatika*), automatons (*automatopoitike*), mechanics (*mechanica*), surveying (*metrica*), geometry, stereometry and the design of theodolites (*dioptra*). He is said to have written a treatise about vaulting (*kamarika*).

Details of tools and their use shall be discussed under »Execution«

2.4. Costs and financing

2.4.1. Costs

Little is known about the cost of roads. The few sources concern repair works. The curator T. Vibius Tomundines spent 150,000 Sesterces (SH) per mile for the repair of 20 miles of the Via Caecilia. Under Hadrian 16 miles repair of the Via Appia cost 1,716,000 SH, amounting to 107,250 SH per mile. Other examples vary between 66,000 and 111,500 SH per mile. Obviously, local conditions had a great influence. The cost of new roads must have been a multiple of the above figures; Pekary (1968) estimates half a million SH per mile.

Costs of stone bridges are easier to establish, since they constitute well defined, singular structures. We do not know how costs were shared by land owners, local communities and the state. As a rule, labor took the major part of expenditure and the kind of workers employed, whether free masons, slaves or soldiers, will have influenced the costs. Slaves did not get paid; they were taken care of by their owners, who rented them out at a price. Neither can we base a calculation in reference to today's costs, but records of bridge construction before the 19th century could help us, since technical and social conditions were similar.

Let us take as example the Roman bridge at Alter do Chão in Portugal (see also case study SP 42). Six equal arches in rusticated masonry represent technical perfection, harmonious proportions and durability – in one word beauty.

Built at the beginning of the 2nd c. AD, it crosses the Seda River as part of the Via Augusta from *Emerita Augusta* (Mérida) to *Olisipo* (Lisbon). 115.3 m long, 6.75 m wide and about 9 m high, it is in very good condition and still handles the full traffic on highway Nr. 369 (last renovation 1950).

Based upon the mass of the bridge and the productivity of the average worker, we estimate a workload of 400,000 man-hours, or 40,000 ten-hour workdays. Given 10 months construction time this translates into a workforce of about 200 men. One work-day cost 8 sesterces = 320,000 SH. Adding one third for materials and transport = 100.000, ergo a total cost of 420.000 SH.

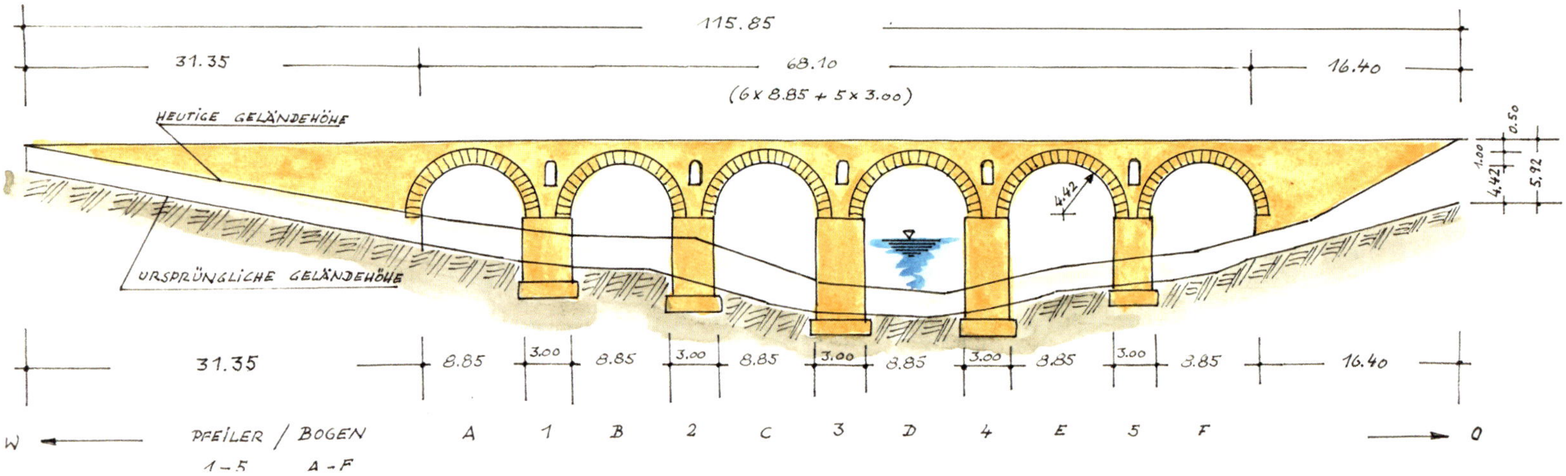

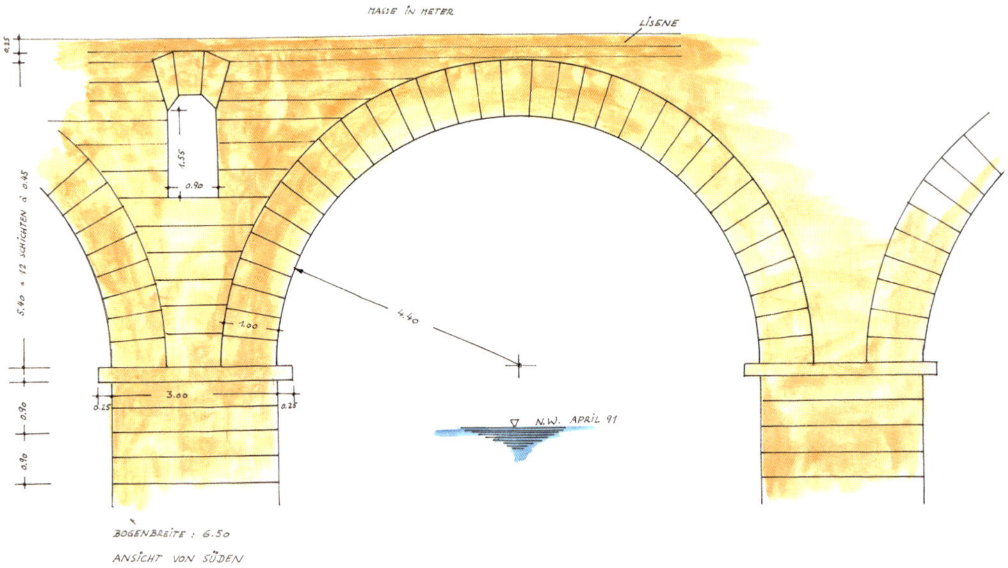

43. Bridge at Alter do Chão, Portugal. View from the south.
44. South elevation.
45. Detail of an arch.

46, 47. Inscription on the Treponte bridge south of Rome along the Via Appia, marking its construction by Trajan.

This seems little in comparison with then prevailing costs as determined by Finlay (1980): A plaque from Velleia near Piacenza sets the price of an estate at 300,000 SH. A daughter of Sulla bought a luxurious villa for 300.000 SH (later sold it to Lucullus for ten million!) The wealth required of a praetor was about 400,000, of a senator 1 million.

On the other hand, K. Christ (2001) assumes 8 SH as average daily cost of a builder in the 1st c. AD, including all extras. For the 2nd c. AD Finley sets the yearly pay of a legionary at 1200 SH, from which costs for food were subtracted.

These are extremely rough calculations and should only indicate that before the big inflation, starting in the 3rd century, construction costs for roads and bridges were limited, mainly by employment of slaves and soldiers and state-owned sources of materials.

In the preamble to the Xth book Vitruvius tells us of good and bad contractors:

»1. It is said that the forefathers of the famous, large Greek town of Ephesus instituted a law, with a tough, but not unjustified proviso. If an architect assumes the supervision of a public construction, he declares how much the building will cost. After the cost estimate has been handed to the authorities, his assets are held as pawn until the work is finished.
Is it finished and the building costs conform to the estimate, he shall receive an honorable citation. If the cost did not exceed more than a quarter of the estimate, this quarter shall be covered by the government, without fine for the architect.
However, shall the additional cost exceed a quarter of the estimate, the required sum shall be taken from the architect's assets.

2. Should only the eternal Gods have seen fit to have the Roman people promulgate such a law, not only for public but also for private construction. Then there would no people run around unpunished, who know nothing of building (but pick up construction jobs). Instead, only people with solid formation and scientific knowledge would dare to become builders. Private people would never have to pay without end and lose their fortune. Afraid of punishment, the builders would prepare the estimates with great care, so that their clients could finish the construction within their budget, or just need a small extra sum.«

2.4.2. Financing

The first roads out of Rome will have been paid by the Republic. Purely military roads to the limits of the empire were built by the soldiers and paid from the military budget.

In imperial times, Pekary (1969) points out, inscriptions on buildings and mile stones often indicate that the rulers themselves paid for the works (*sua pecunia*), Hadrian even specifies the amounts. But Pekary also holds that the emperors' engagement as paymasters was the exception and is exaggerated by the preserved inscriptions. He thinks, most of the costs were absorbed by the land owners (*possessores agrorum*) served by the road, in particular those directly adjacent to it. They also had to maintain it. The rate of participation depended on the size and productivity of their property, which was the basis of taxation. Of course, this entailed many controversies, in particular when communities were scattered in uneven territory.

Suetonius (AD 75–150) reports that successful generals, who were honored by a triumph (*triumphalibus viris*), were assigned road sections for which they had to bear the costs out of their spoils of war (*ex manubiali pecunia*).

From an application, which was required for the placing of a mile stone, we learn that emperor Maximinus Thrax (AD 236–238) ordered repair works in many provinces. He denoted this with inscriptions such as *vias et pontes vestutate, collapse restituit* or *pontes refectum et vias munierunt et milliaria restituerunt* (dilapidated roads and bridges were repaired, collapsed bridges rebuilt, roads secured and milestones replaced) (CIL III).

On the triumphal arch of the Alcántara Bridge over the Tejo River (Spain), built at the time of Trajan, an inscription reads: »*municipia provinciae Lusitaniae stipe conlata quae opus pontis perfecerunt …*« (this bridge was completed with the taxes of the free cities of the province of Lusitania, Portugal), followed by a list of the cities contributing financially to its construction – although the bridge was not on their territory. (CIL II)

The costs for secondary roads had to be born by the communities concerned. They were often picked up by private individuals, as recorded in inscriptions (CIL V, X)

2.5. Contracting and execution

2.5.1. Contracting

Building contracts (*lex locationis*) were normally made in writing. The works were described in detail, setting the amount and quality of materials, prices, deadlines, number and boarding of workers, contract penalties, guarantees, arbitrage and modalities of payment. The purpose of such contracting was the selection of the most suitable bidder. In imperial times the emperor or the *duumviri* awarded contracts, being counseled by a *consilium*. The contractors had to furnish bail and bond money to assure quality, completion dates and prices.

2.5.2. Execution

2.5.2.1. Site preparation and workforce

After clearing the site and the preparation of access roads, large, horizontal working surfaces were prepared, to facilitate the moving of materials, to support lifting devices and to serve for the lay-out of plans and details. The working drawing of an arch at full size has been found in front of the theatre in Capua (2nd c. BC) and on a plastered wall of the theatre lobby (*parodos*) at Terracina (1st c. BC). In Baalbek (Lebanon) on the podium of the Temple of Jupiter, the whole pediment design is engraved in elevation. Similar lay-outs exist on the exterior wall of the Temple of Apollo at Didyma (Turkey). They are scratched into red painted rendering, resulting in drawings with sharp white lines. (Haselberger, 1984). Haselberger also identified the well known drawing in the paving of the Mausoleum of Augustus, as draft of the Pantheon's portico (Haselberger, 1994). Although at a distance of 800 m, but close to the Tiber, it was an ideal working surface for the preparation of the portico's masonry.

In peace time, and mostly in the provinces, the Legions were frequently employed for the construction of roads and bridges. »Those detachments (*vexillatio*) are for special tasks in- or outside the governor's province, e.g. for occupation or base duties. They also have to build roads, bridges or fortifications and are to prepare and to transport building materials.« (RE VIII A)

Near the end of the Republic the legions adopted special detachments for road construction (*centuria accessorum velatorum*), a kind of military sappers under an officer's command (Forbes, 1934). They had civil engineers as technical staff, whose names are often cited in dedications (Cüppers, 1969).

A 1st c. AD inscription, found on the western abutment of the bridge from Mainz to Kastel over the Rhine, reads: »Of the 14th Twin-Legion (*Gemina*), the victorious Martian, the century of Quintus Blandius Fanester (was engaged in the construction of this bridge).« (BRKG 40, 1959) The soldiers did not like such assignments and at the time of excessive road construction under Augustus there were some riots. If the military was not responsible for the works, the projects were given to contractors (*redemptores, conductors*). With time small contractors merged into large, industrially organized building companies. They executed large public works, by engaging free artisans, freed slaves, slaves or prisoners of war.

The artisans were organized in guilds (*collegia*), controlled by the government. In the imperial correspondence of *Plinius Secundus* (letter 34), Trajan warns, that the guilds could in time become secret associations.

48. Legionaries constructing a building, from Trajan's column in Rome.
49. Construction yard at the mausoleum of Augustus, early 2nd c. AD. In front, shop drawing for the Pantheon's portico (Haselberger in *Antike Welt*, 4).

Among the artisans the stone masons (*lapidarii*) were the most prominent and best paid. Work was paid per finished piece, rough preparation was done by less skilled workers. Other specialists were carpenters (*carpentarii*), blacksmiths (*aerari, ferrarii*), surveyors (*mensores*) and scaffolders. The possession of slaves corresponds to today's equipment and machinery of an enterprise.

There is little knowledge about the formation of artisans in construction. It generally was the apprenticeship system: From apprentice to journeyman to master. Often the skill and experience of a trade was passed on from father to son through generations. Some of them would rise to leading positions in government (*praefecti fabrum*). Vitruvius is said to have served Caesar and Augustus in this function.

Legionaries had many possibilities to acquire building skills, by building camps, roads, bridges, canals and war engines. Local workers would be familiar with available materials, the handling of granite, limestone, marble and tuff. They would know the different kinds of timber and the best time of lumbering. In imperial times whole teams of specialists travelled from site to site, according to exact schedules. Without such organization the execution of the many large projects would not have been possible. The construction teams generally consisted of ten men (*decuriae*) under one supervisor (*exactores operum*), who checked the quality of materials and workmanship. He was the equivalent of today's foreman.

In charge of the works could be an architect/engineer, a military engineer, the contractor, an experienced slave, a »guest worker« (often from Greece) or an official (*curator viarum*).

Contracting road works was big business and we know of several cases of corruption. Tacitus reports in his annals (III, 31): »Corbulo (previously praetor at the time of Tiberius, AD 14–37), complains bitterly that many roads in Italy are in very bad condition, due to crooked contractors and lack of supervision by government officials. He liked to bring such cases to court, but not so much for public benefit, since he prosecuted the culprits so relentlessly that he ruined their reputation and finances.« According to Dio Cassius (XIL, 15, 3) Corbulo went so far, that Emperor Claudius (AD 41–54) forced him to pay back some confiscated money from his own pocket.

2.5.2.2. Construction time

Records and inscriptions give us some idea of construction times. The 6th c. AD chronicler Prokopius of Kaisareia (Palestine) mentions in his book *De aedificiis* (VI, 5–6) a report by Apollodorus of Damascus, stating a construction time of one year for his wooden Danube bridge at Turnu Severin (1100 m over 20 piers). This was considered a construction miracle. Cassius Dio (163–229) gives the dimensions of the stone piers, which translates into about 150,000 m^3 of masonry work, needing 1300 masons for 10 months. Adding carpenters for the timber arches we reach a specialized workforce of about 3500 men. Given the enormous supply and logistics problems we may assume a total of 10- to 15,000 workers being engaged in this project.

Caesar also used timber for his 400 m bridge over the Rhine and claims only 10 days for its construction (*Bellum Gallicum*, IV). On the other hand, when the enemy could not be engaged, it took 18 days to dismantle the bridge after his retreat. It would have been easier to use ships for crossing the river, but Caesar declares this »below his and the Roman people's dignity«. Thus the bridge construction was above all meant to demonstrate Roman power.

C. Julius Lacer mentions 2 years as construction time for the 200 m stone bridge over the Tejo at Alcántara (Spain). It is 50 m high and must have required heavy piers.

If these construction times seem rather short, we have to consider Roman talent for organization. Surely, the project of a bridge was well prepared, defining all the technical details involved. This enabled the work to proceed simultaneously on foundations, quarrying and cutting stones, readying access roads and the like. As said above, the availability of travelling specialist teams also helped a lot, as did the commandeering power of the government.

2.5.3. Carrying out the work

Trajan's column beautifully depicts legionaries laying ashlar masonry, as well as filling the space within the facing stones of piers with concrete. Formwork and stone finish were combined, resulting in big savings of time and material. Indeed, this technique enabled the Romans to build huge structures such as the Colosseum in a fraction of the time expended on a much smaller temple, which was built in trabeated design to suit tradition.

2.5.3.1. Lifting of stone blocks

Light materials like brick, small stones, mortar, metal parts and wood could be manually handled with the help of backpacks, stretchers, buckets and boxes, or by bucket chain. When the weight of a single component surpassed 50 kg (an ashlar of 75 x 50 x 50 cm weighs about 400 kg), moving and lifting devices were needed. For horizontal or slightly inclined movement rollers on a smooth surface were used. Most of the time, the building units had to be lifted to the required position. A windlass was the simplest implement, for hoisting- winches, which need a metallic gear to connect with the crank, were not used.

To increase the hoisting capacity, pulleys had to be employed. They were attached to pulley blocks or trestle cranes, which were fixed to the ground and held up with staying cables.

Vitruvius (X, 2) tells us under hoisting machines: »Two timbers are provided, strong enough for the weight of the load. They are fastened together at the upper end by a bolt, then spread apart at the bottom, being kept upright by ropes attached at the upper ends and fixed at intervals all around. At the top is fastened a block, which some call *rechamus,* it has two sheaves turning on axles. The traction rope is carried over the sheave at the top, then let fall and passed around a sheave in a block below. This is brought back to a sheave at the bottom of the upper block, and so it goes down to the lower block, where it is fastened through a hole in that block. The other end of the rope is brought back and down between the legs of the machine.«

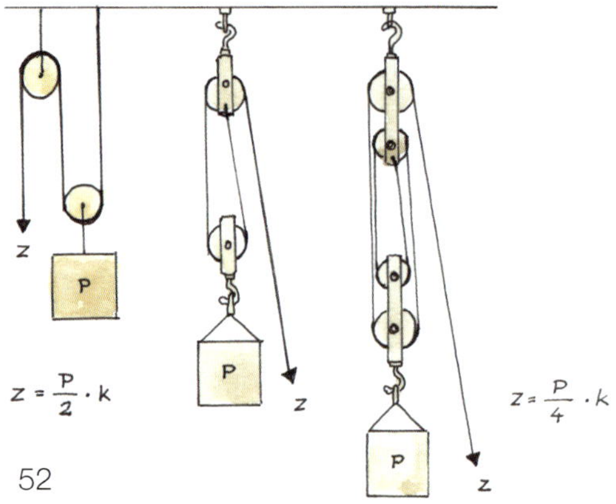

52

53

50

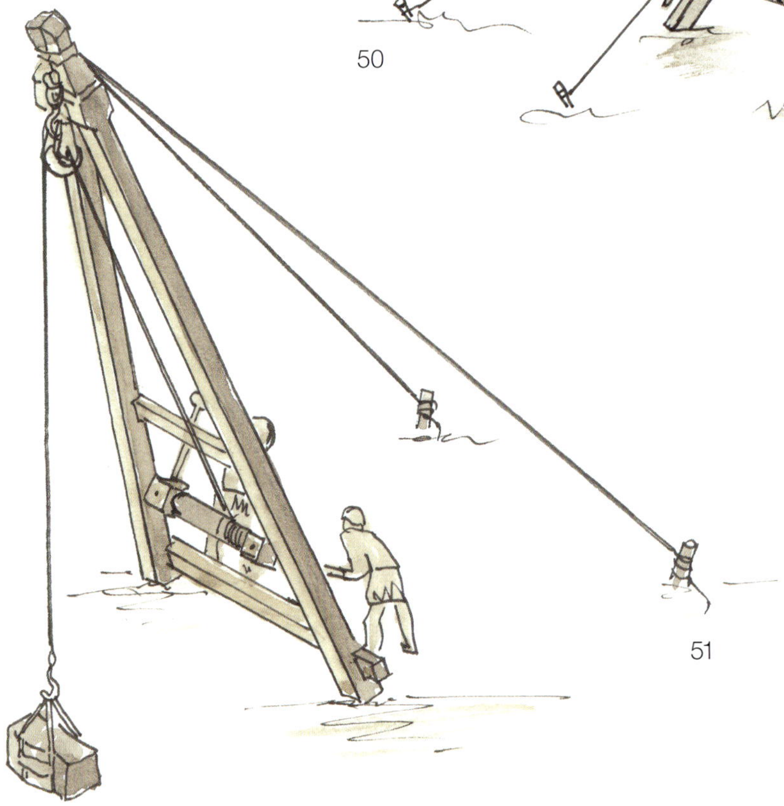
51

50. Hoist with a three-pulley *tripastos* (1:3 advantage).
51. Simple hoist on a two-leg trestle with staying cables and windlass operation.
52. Operation of pulleys. A and B are pulleys reducing the pulling force Z to half the load P times friction losses k. (A = simple pulley, B = compound pulley, C = block and tackle pulley system reducing pulling force Z to a quarter of the load, P, times k).
53. Hand-operated windlass. If we assume 40 cm long grab handles for leverage, a 10 cm diameter spindle, 15 kp pulling force of a man and allow some loss for friction, one man could lift about 50 kp. Several workers will multiply the effect.

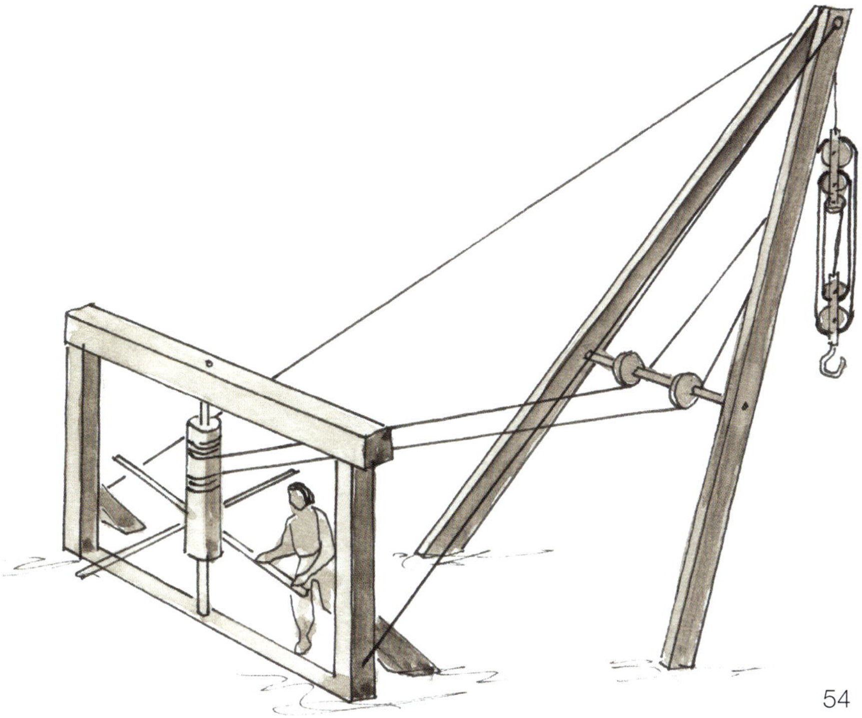

54

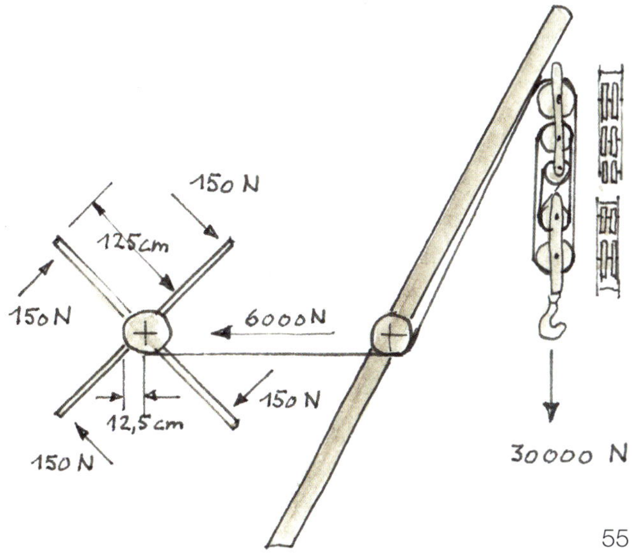

55

54. Trestle crane with *pentaspastos* (1:5 pulley) and capstan operation.
55. Operation of the capstan. To lift very heavy loads the treadmill was employed.
56. Heavy two-leg trestle with treadmill.
57. Drawing of a relief in the Capua theatre of a treadmill-operated crane used by the column manufacturer Lucceius Peculiaris.

57

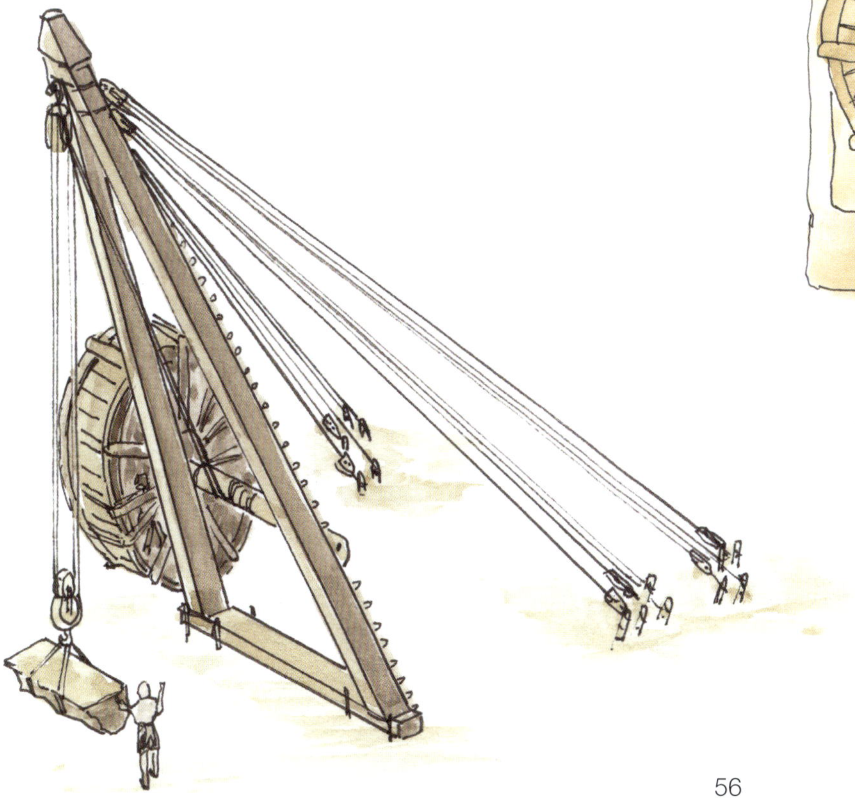

56

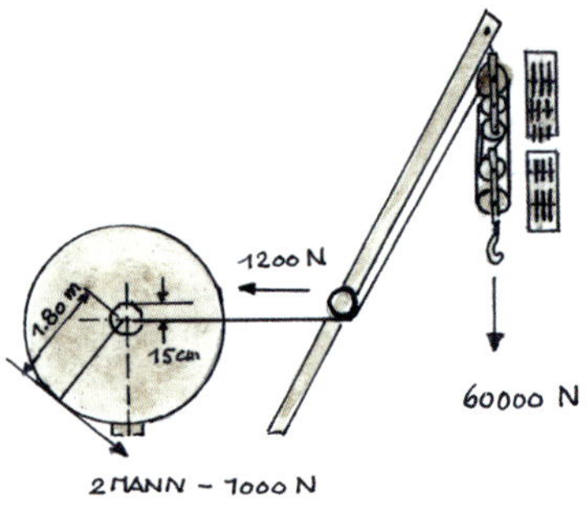

58

59

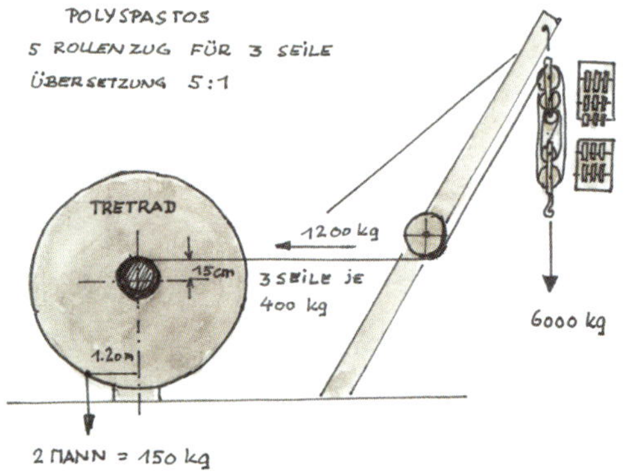

60

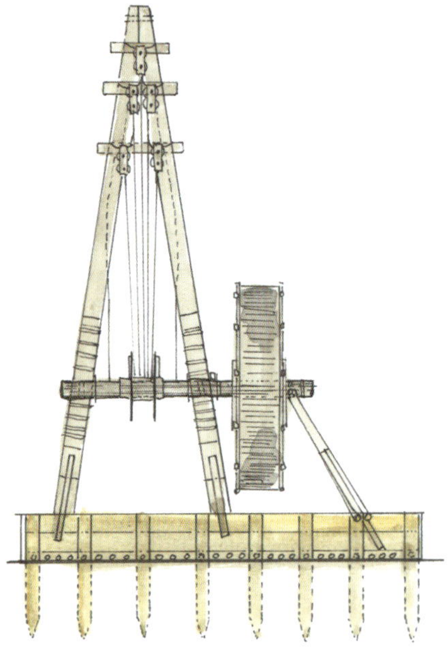

61

58–60. Polypastos with a two-men treadmill (Kretschmer, 1983).
61, 62. Reconstruction drawings of a crane with treadmill.
63. Relief from the tomb of the Haterier family (1st to 2nd c. AD) in the Vatican museum, showing a crane with treadmill.
64. Reconstruction of the »Hatarier crane«.
65. The Roman derrick was a lifting device less strong, but more flexible in positioning the load. It consisted of a vertical mast with a boom extended from a collar (Landels, 1989).

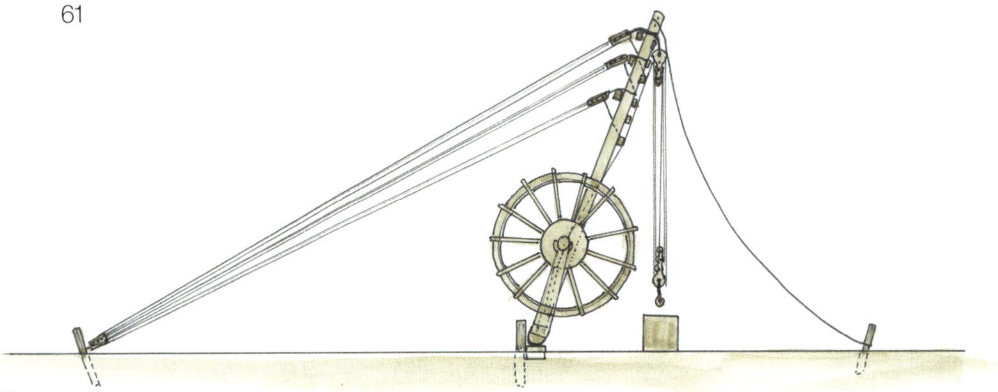

62

63

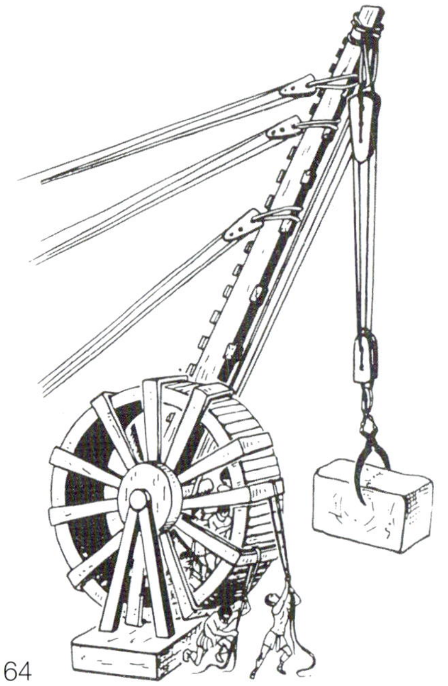

64

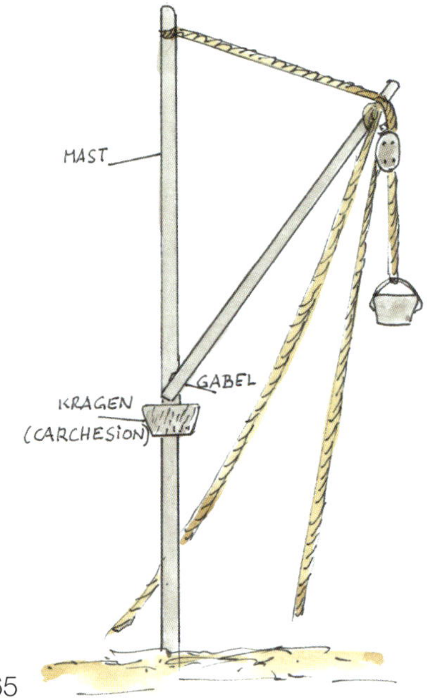

65

66. Straps attached to knobs. Disadvantage: more stone cutting to produce and remove the knobs.
67. U-grooves to receive rope loops. Ropes are easy to retrieve.
68, 69, Lifting of stones by means of lewis bolts.
70, 71. Variation of the lewis method, employing a scissor hoist.
72, 73. Variations of lewis bolts with different plugs and feathers. A dovetailed hole had to be produced, about 12 cm long by 3 cm wide and 12 cm deep with 5° slant. The iron hooks were secured by inserts. Problems: difficult cutting and a severe wedging force when lifting, requiring strong stone material.
74, 75. Chain dog. The scissor action had limited lifting capacity, but required little work.

2.5.3.2. Placing of stones

How can we get a hold on a heavy, dressed stone, lift it, place it and retrieve the lifting device? To solve this problem the Romans used four methods (Adam, 1984, Landeis 1989, Lugli, 1957, White, 1984):

1. projecting knobs to attach a rope,
2. grooves to receive a rope,
3. lewis bolts or scissor clamps,
4. chain dogs.

66

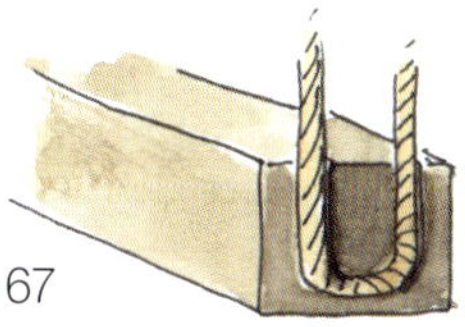
67

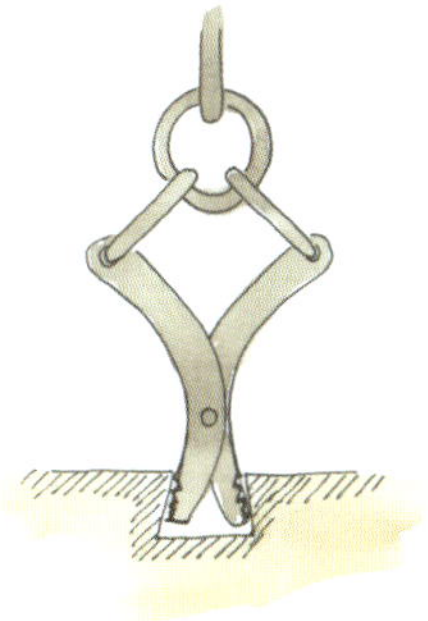
70

71

74

75

68

72

69

73

For the horizontal positioning of blocks crowbars were used to move the ashlars horizontally. Small holes in the surface of the range below served as a grip. A lime slurry acted as lubricant but did not leave a mortar joint.

76, 77. Examples of lifting devices shown by Piranesi in *urbis aeterna vestigial* (1786).
78. Use of crowbar and wood block for shifting of stone block (Adam, 1984).
79. Lateral shifting of stone from a scaffold (Adam, 1984).
80. Undercuttings used for the insertion of crowbars (Müller, 1980).

81. Compacting Roman concrete as foundation for a wall (Jocundus, 1511).
82. Compacting Roman concrete between the formwork of a wall (Adam, 1984).
83. Wooden slip form system for a 35 cm concrete wall, poured in 150 cm high increments. (Kantholz = square timber, Zange = tie, Schalbrett = shuttering, Abstandshölzer = distance bars, frisch eingebauter Gußbeton = fresh concrete, Gußbeton des vorhergehenden Abschnitts = previously cast wall portion, Betonierabschnitt H 130 cm = pouring height 130 cm).

2.5.3.3. Concrete work

Till today, wooden boarding serves as formwork to receive fresh concrete. Care has to be taken that the formwork can withstand the pressure of the heavy material in its plastic state. Working height was around 1 to 1.3 m. The thickness of a wall was given by inserted spacers; the horizontal pressure was taken up by anchors in the masonry below and tie bars or ropes above the forms.

Wooden tie bars could be carried through the finished masonry, projecting on both sides to serve as support for working platforms. When the concrete had set they were either removed or cut off. The top bar could be used as support for the next layer's platform and the formwork could be re-used, anticipating today's slip form system.

Very frequently, and particularly in bridge construction, the exterior stone masonry served as formwork, having internally projecting tie stones to bond with the concrete. Very widely used were shells of triangular or pointed bricks, their projections bonding well with the concrete (*opus reticulatum*).

Roman concrete was not poured, but built up in alternating layers of stone rubble and cement mortar, which was compacted with rams to fill all voids. Workers carried stones and mortar in baskets or buckets to the place of application.

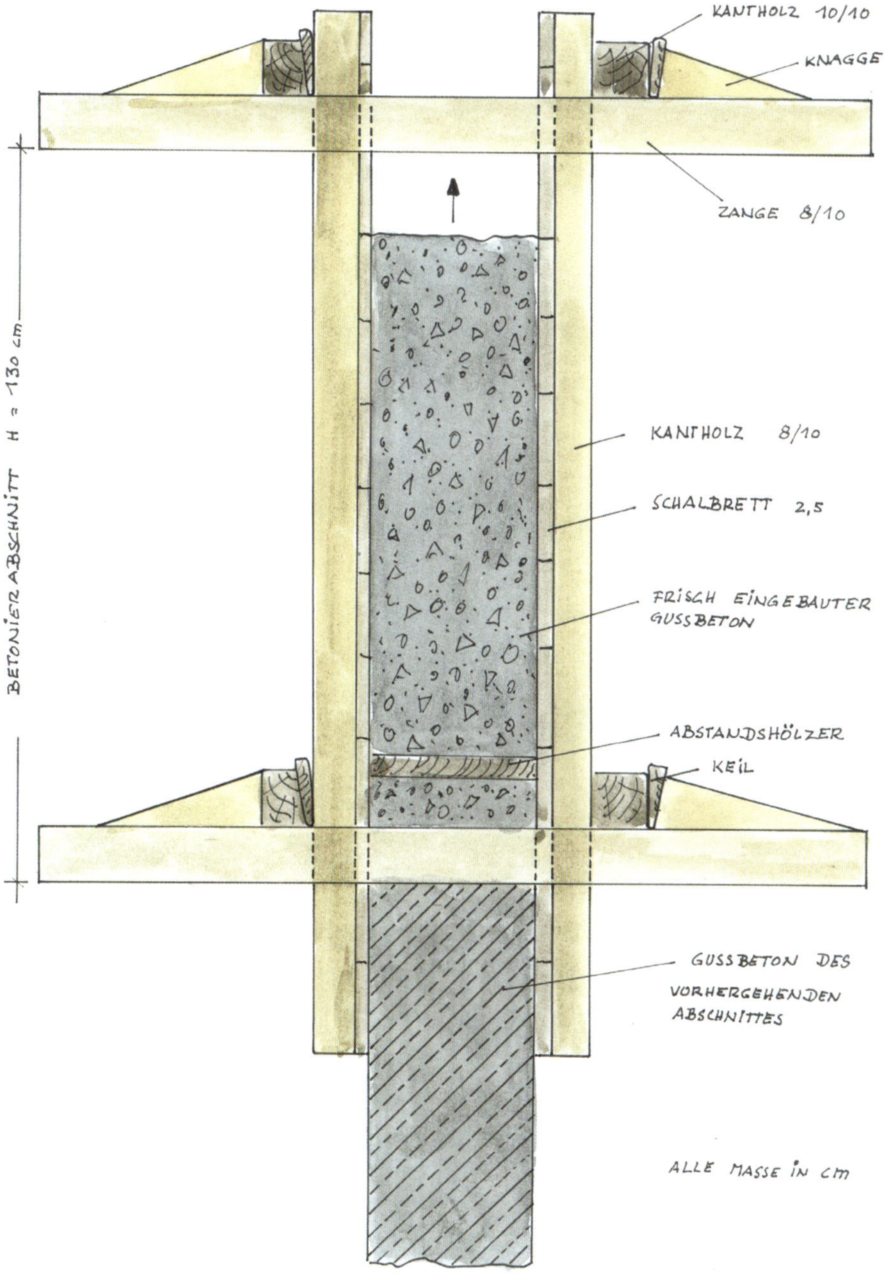

2.5.3.4. Scaffolding

Scaffoldings are self-supporting temporary structures to facilitate the building process. They range from the trestle support to multi-level frames and are composed of timber poles, cross-bars and diagonal braces plus boards as work platforms. The wooden members are connected with ropes, nails or cleats, to provide for easy dismantling and re-use. Today's methods are very much the same, although the parts are of steel.

84. Trestle scaffold.
85. Twin-poles scaffold with diagonal bracing along a wall.
86. Single-pole scaffold attached to a wall.
87. Cantilevering work platforms (Adam, 1984).
88. Pole connections.

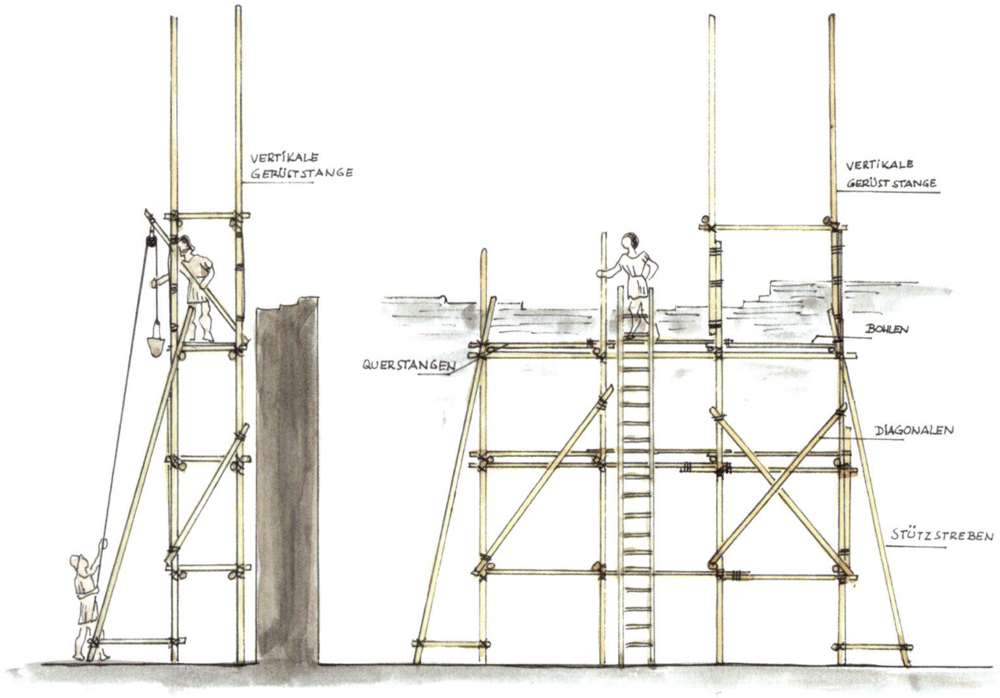

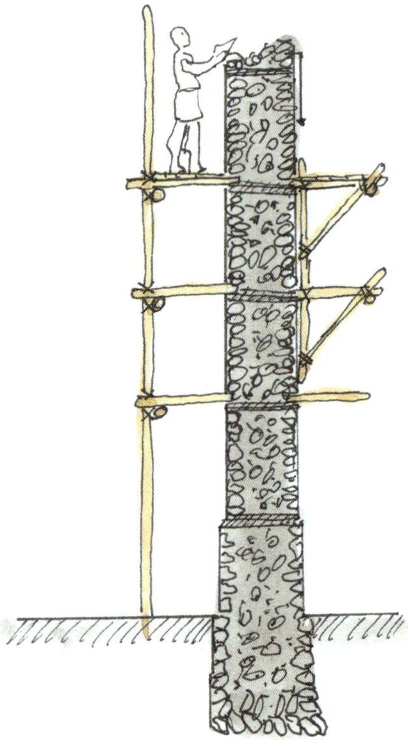

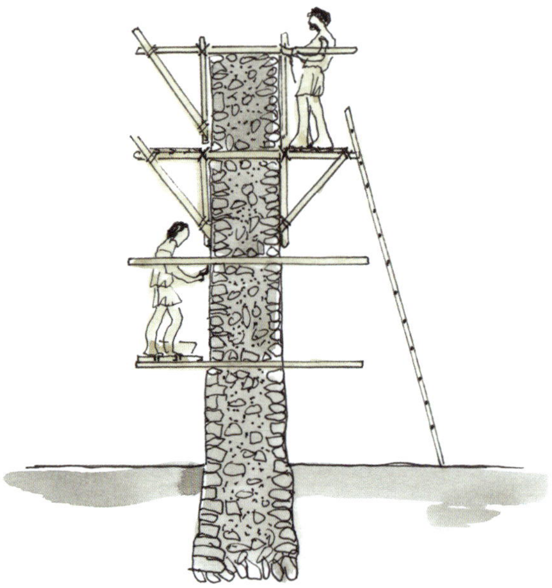

2.5.3.5. Falsework

A temporary support is needed to place the voussoirs of a true arch until it is complete. There are practically no Roman references, but we may take pre-modern examples for information, since technology had hardly changed. Timber was used and to economize, the falsework was started as high as possible, from the abutment up. For this purpose projecting consoles were provided in the masonry of the piers.

An arch of 10 m span and 5 m width – we really should call it a vault – had a weight of 150 tons, discounting the portion carried by the abutment, before the key stone was placed. For the temporary support of this arch about 20 to 40 m^2 high quality timber would be needed.

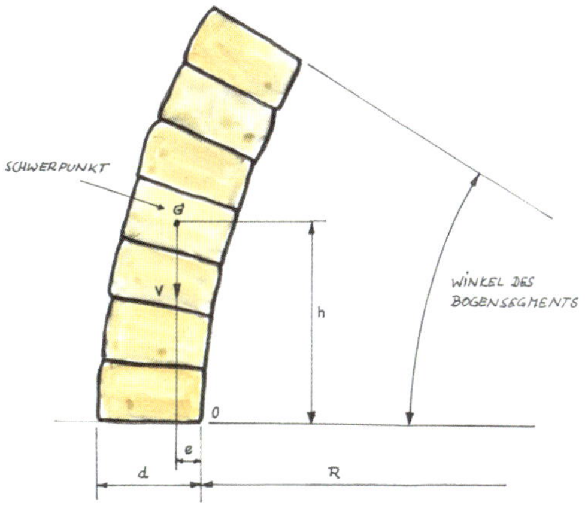

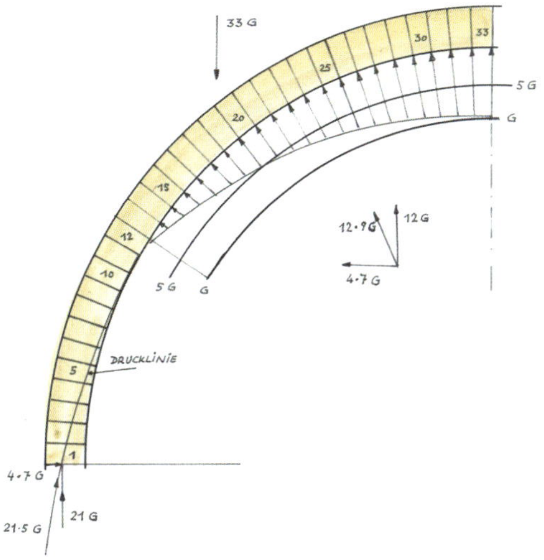

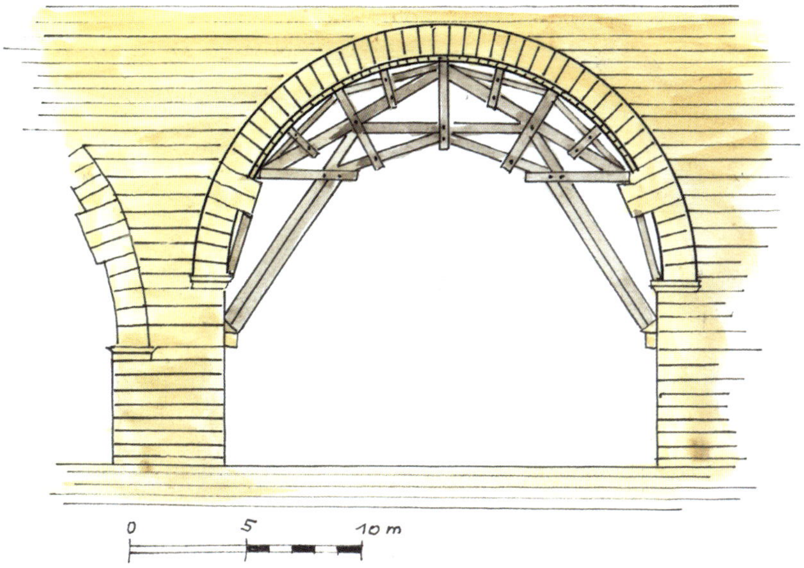

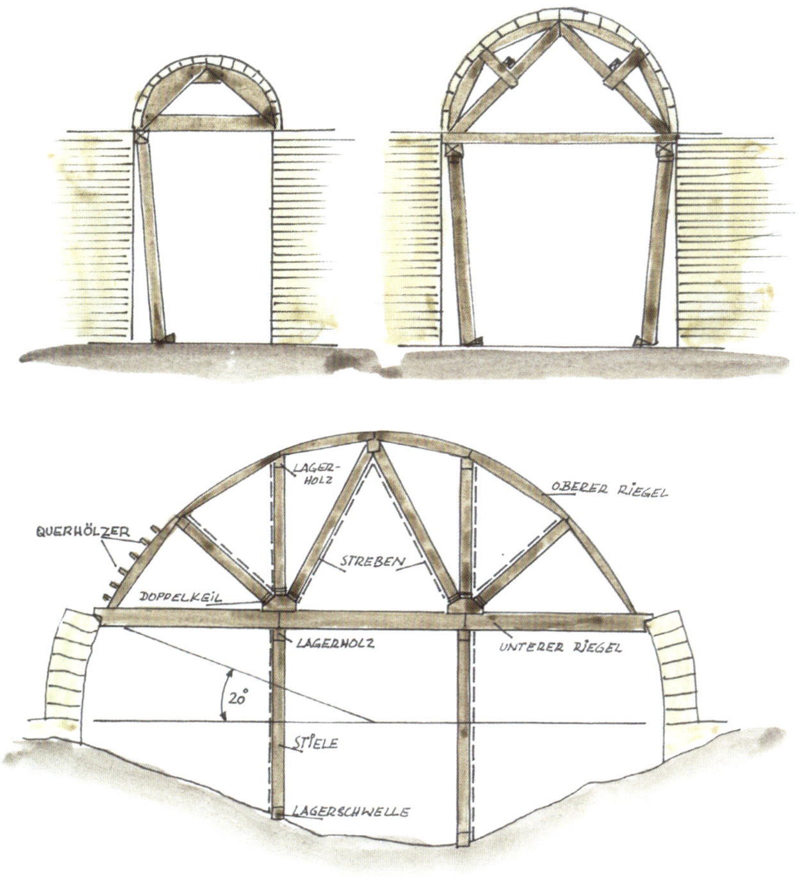

89. Portion of arch carried directly by abutment. (Schwerpunkt = center of gravity, Winkel des Bodensegments = up to 30° arch segment).
90. Diagram of falsework loading under an arch without keystone; the lower curve with radial arrows indicates the increasing load on the formwork (O'Connor, 1993).
91. Falsework at the Pont du Gard, sitting on pier consoles (Adam, 1984).
92. Three magnitudes of falsework for arches.

To further simplify the work, stone slabs or lighter brick could be put as »lost falsework« for individual arch »ribs«. The centering then was shifted inward and in several small steps a vault was produced, to receive the real arch construction. 17th and 18th century illustrations show the construction of bridges, which was similar to the Roman methods. On bridges with repetitive arches the Romans will surely have used a sequential method, reusing centering and cranes as they went along.

As there is substantial initial settlement, due to many joints and the great weight, the falsework had to receive a camber to achieve the correct semicircular arch shape in the end. The removal of centering was a very delicate procedure. In order not to introduce imbalances, it had to be done by uniformly loosening the wedges below the centering, all along the springing of the vault.

93. Detail of the Pont du Gard.
94. Arch composed of five stone ribs and concrete.
95. Stone and brick faced concrete bridge (Ponte del Pondel, Val di Cogne, Aosta, Italy).
96. Sequential method of multiple arch construction.

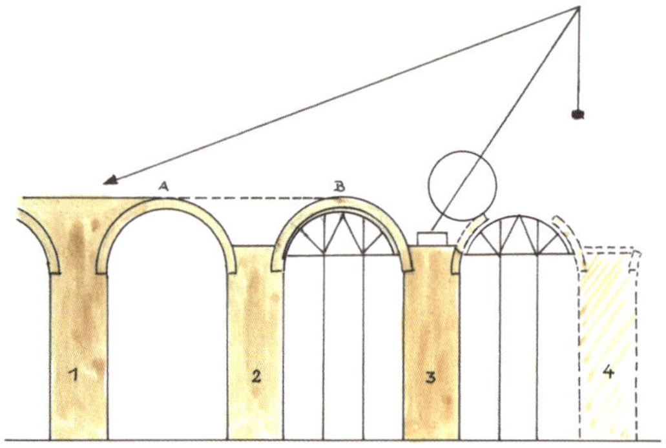

97. Section through a water well. (Undurchlässiges Bruchsteinmauerwerk = watertight rubble masonry, durchlässiges Trockenmauerwerk = pervious dry masonry).
98. Simple pile driving platform.
99. Floating pile driving platform as suggested by Köthe (1991).

2.5.3.6. Piling

Piling has to be used for very poor soil conditions. Vitruvius (III, 4) says: »If solid ground cannot be found, but the place proves to be nothing but a heap of loose earth to the very bottom, or a marsh, then it must be dug up and cleared out and set with piles made of charred alder or olive wood or oak, and these must be driven down by machinery, very closely together like bridge-piles, and the intervals between them filled with charcoal, and finally the foundations are to be laid on them in the most solid form of construction.«

The Romans used floating ramming devices to drive iron pointed piles into riverbeds. They consisted of up to 20 m high trestles, holding the pile in position and having a guiderail for the piling hammer. For quick operation the hammer of about 500 kg was pulled up over a cable roll by a crew and then released for free fall. Piles should be driven in short intervals to avoid their getting stuck in the ground. Fortunately, as the pile descended into the ground and gathered strength, the driving height and power of the hammer also grew. Rarely were winches employed for ramming.

2.5.4. Excursus: Water lifting devices

When digging wells or excavating for foundations below the water table, sheet piling or cofferdams are needed to keep the water off. This entailed the removal of water seeping into the work space. Of course, more often water was lifted for domestic or agricultural purposes. Hand carried buckets or rope and bucket had their limits. The Egyptian *shaduf* was called by the Romans *ciconia*, which aptly means stork. It counterbalances the weight of the water and eases the lifting effort.

The bucket chain is an efficient device to lift water out of limited space. It can be driven by animals or operated with a treadmill.

Water lifting wheels needed much space and lifting height was limited by their diameter. Vitruvius discusses the *tympanum* (X, 4) with eight compartments and water wheels with small buckets, discharging the water into a higher channel. They would be operated by hand or like a treadmill. Famous are the *noria*, river driven waterwheels in Hama (Syria) with up to 20 m in diameter.

100. Eight-compartment *tympanum* as described by Vitruvius. (Gerinne = conduit).
101, 102. Man-driven bucket wheels (Wölfel, 1990.)
103. River-driven bucket wheel after Vitruv (Wölfel, 1990).
104, 105. Crank-operated bucket wheel (104); river-driven bucket wheel (reconstructed after Vitruvius X, 4–6 by Barbaro 1556).

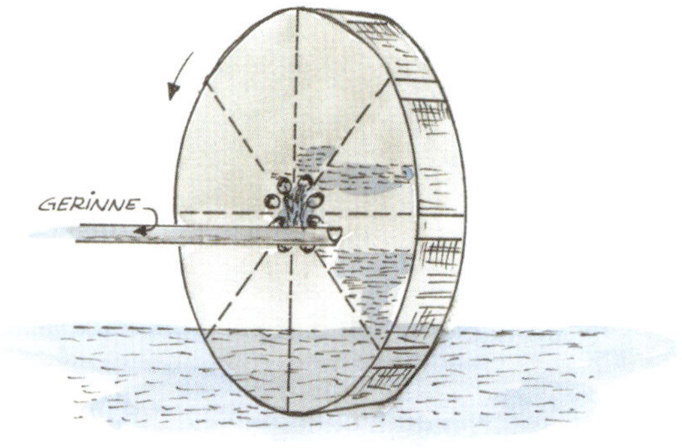

106. View and section of a Roman bronze pump (Wölfel, 1990).
107. The screw of Archimedes.
108. Pump of Ctesibius. (Einlaßventil = intake valve/flat disk, Auslassventil = discharge valve/ flip gate).
109. Roman single chamber pump found on a commercial ship (Landels, 1989).

Mechanical devices applying physical laws were the forte of Archimedes. He devised the water screw (Vitruvius, X, 6). Landels figures for 1,16 m lifting height at 40 % efficiency a lifting volume of 160 liter per minute, or 10,000 l/h. Pumps were described by Vitruvius (X, 7) and Heron of Alexandria. They mention the »machine of Ctesibius«, a two-chamber pump which could lift about 800 liter per hour 5 m high.

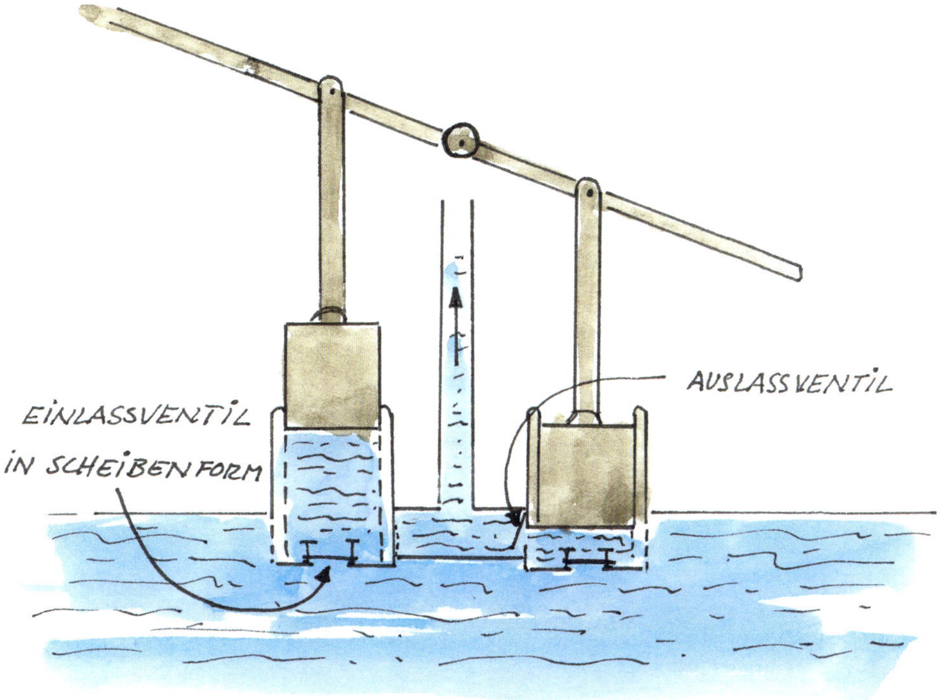

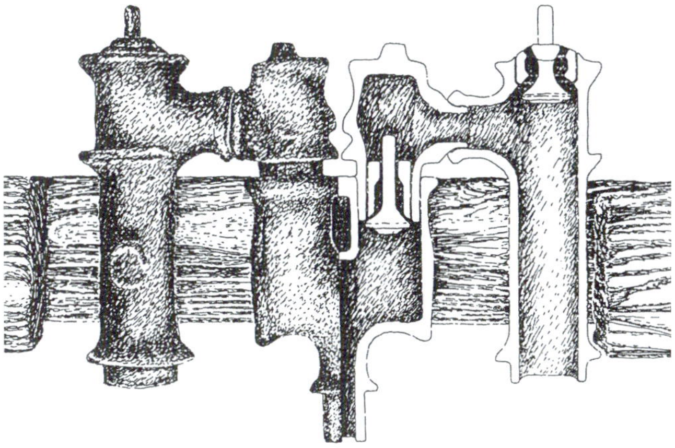

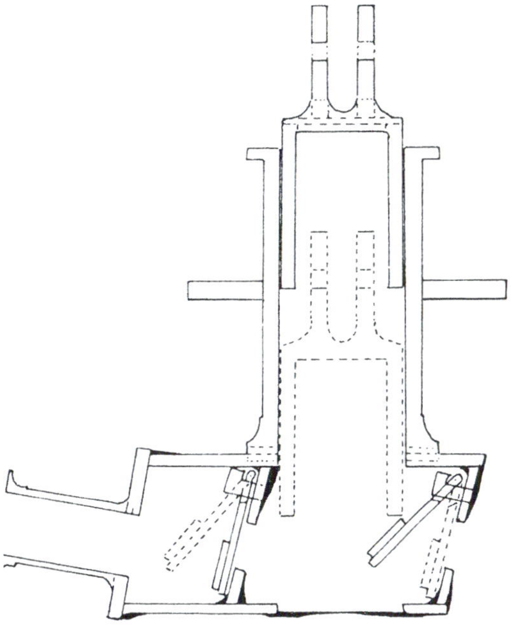

3. The Roman road network

110. Roads radiating from Rome.

3.1. Evolution of the Roman road network

Since the Middle Ages we say »All roads lead to Rome«, but when the Romans created their empire the reverse was correct: »All roads issued from the city of Rome.« 23 major roads fanned out from the city, like the spikes of a wheel.

This was emphasized by the »golden mile stone« (*miliarium aureum*) in the Forum, placed by Augustus in 20 BC. It was the symbolic »Zero Mile«, with distances to the major cities of the empire engraved in gold. The Roman mile = 1475 kilometers derives from milia passum, meaning thousand (double) steps.

The teachers of the Romans were the Etruscans, who had an extensive network of roads as early as the 5th c. BC. The Etruscan city of Marzabotto near Bologna had four 15 m wide streets with sidewalks and drainage. In the beginning, the Romans were occupied with conquering Italy and had little time for building roads, but eventually they built roads in preparation of conquest, or to secure what they had conquered. The growth of the empire can be seen by dating each road. They were originally for military use, facilitating communication, and thus organization and administration. The Romans understood that mobility was the key to success and road construction is among the most impressive achievements of the Romans. Extent and power of the empire are expressed by the dense network of roads, as already stated by Dionysos of Halicarnassus (*Antiquitates Romanae,* III 76,5).

Public construction was also a matter of representation; roads and bridges were not only technically perfect, but of designs expressing ever-lastingness, monumentality and power. The grandeur of the Empire was often expressed by triumphal arches, being built at the beginning of roads (Roth, 1990).

Curves and serpentines were avoided. A straight alignment, achieved by cut and fill through hills and valleys, conveyed invincibility. Roads made the *cursus publicus* possible, an efficient governmental communication system. On the other hand, good roads could also be dangerous, when they allowed invaders to advance quickly.

While the bulk of commercial exchange of goods was carried by ships all around the Mediterranean, with time, roads also served for the transport of general merchandise, either by beasts of burden or by carriage. In particular they were needed for access to mines and quarries.

We distinguish the following categories of roads: Major state highways (*viae publicae*), military roads (*viae militares*), city streets (*viae urbicae*), village streets (*viae vincinales*) and private roads (*viae privatae*). The divisions are not always clear, but we have good definitions of public highways by Praetor Ulpius (2nd c. AD) and the technical writer (*gromatici*) Silius Flaccus. Accordingly, a *via publica* had to serve public traffic, was outside of cities on public land, its width was prescribed by a government official, it was maintained by an official (*curatores*) or a contractor (*redemptores*), and it carried the name of its builder. In the Law of the Twelve Tables (450 BC) the minimal width of roads was set at 8 feet or 2.4 m, allowing only single lane traffic. In practice, the driving surface had a width of 4 to 7 m in Italy, or 5 to 6 m in Gaul, bordered by 3 m side strips, giving the roads a total width of 10 to 13 m.

The measure of all things was the Roman mile (*milia passum*) at 1,481.50 m. Each mile stone had inscribed the name and rank of its originator and assorted distances to places of interest (crossings, villages, inns). Mile stones were public documents, in imperial times the governor had to obtain imperial permission for placing a mile stone.

According to the *Itinarium Antonini Augusti*, there were 53,000 miles (80,000 km) of major roads at the time of Caracalla. At the height of Roman power there were 90,000 km principal roads; including secondary roads there was a network of about 300,000 km. At Trajan's time Italy had 16,000 km principal roads. At a population of 8 million this means 2 km per 1000 persons, comparing with today's German equivalent of 2,6 km per 1000 inhabitants.

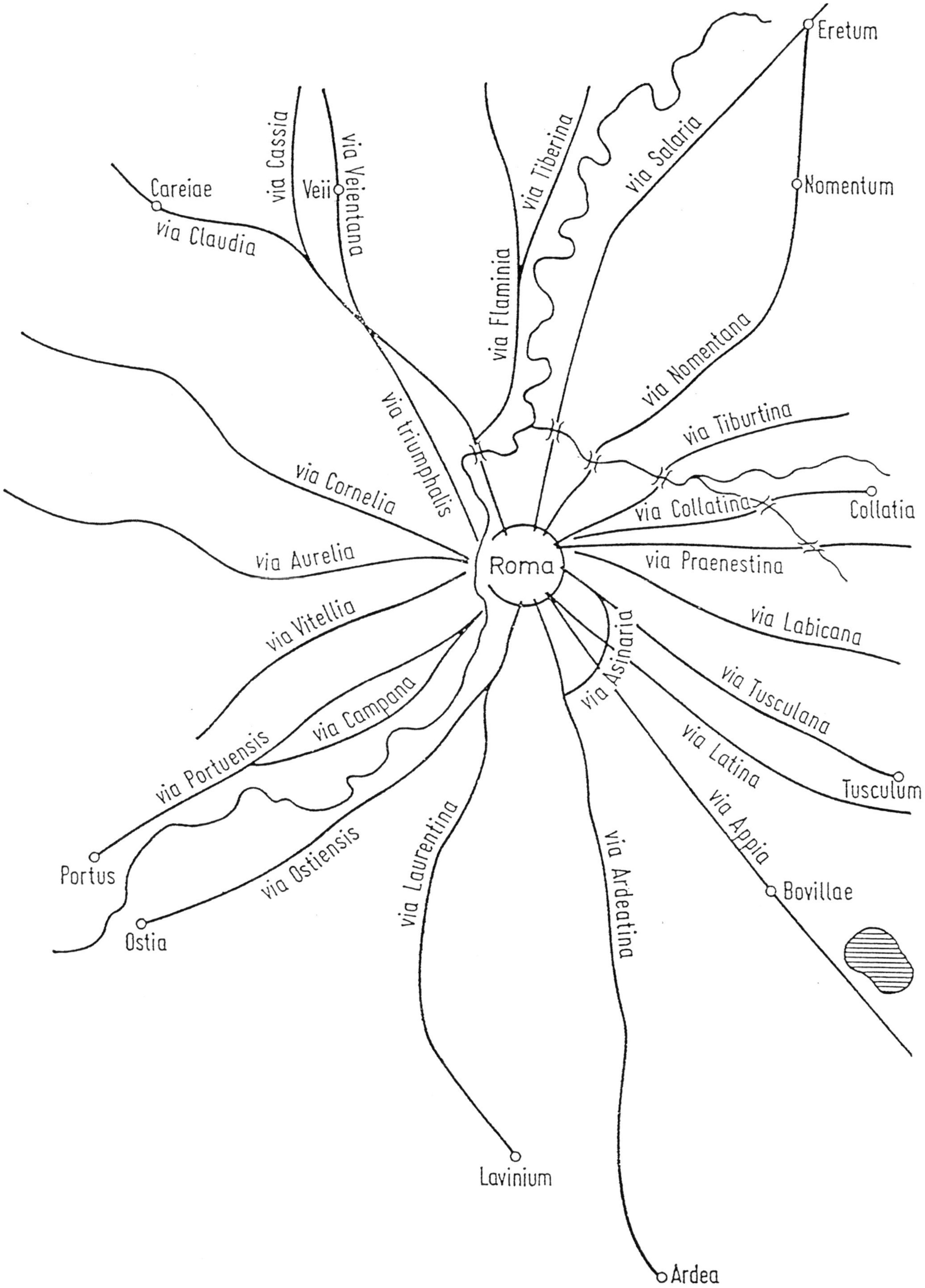
Eretum
Nomentum
Careiae
Veii
via Cassia
via Veientana
via Claudia
via Tiberina
via Salaria
via Flaminia
via Nomentana
via triumphalis
via Tiburtina
via Cornelia
via Collatina
Collatia
via Aurelia
Roma
via Praenestina
via Vitellia
via Labicana
via Asinaria
via Campana
via Tusculana
via Portuensis
via Latina
Tusculum
via Appia
Portus
via Ostiensis
via Laurentina
via Ardeatina
Bovillae
Ostia
Lavinium
Ardea

3.1.2. The beginning of Roman road building in the 4th c. BC

The first famous road was the Via Appia, known as *regina viarum*, or »Queen of Roads«. It was built in 312 BC by the blind censor Appius Claudius Caecus (Livius, 9.29). First, it went unpaved to Capua; in the middle of the 3rd c. BC it was extended to Brindisi, the principal harbor of the southern Adriatic Sea. Initially built for strategic reasons, it soon became the most important road to southern Italy.

Passing through the Pontine Marshes, it was regularly flooded in winter time. A canal parallel to it served for travel by boat. Under Trajan (AD 98–117) the swamps were largely drained, making possible a permanent road with bridges.

When in 232 BC Roman citizens settled in *ager gallicus* near Rimini, the area had to be connected to Rome, which was done by the Via Flaminia. It went over Narni, Carsulae, Acquasparte, Massa Martanaa, Bevegna, Nocera Umbra, Gualdo Tadino, Scheggia, Fano and Pesaro till *Ariminum*, today's Rimini on the Adriatic coast. During the empire an additional branch, the Via Flaminia Nuova was added, from Narni over Terni, Spoleto, Trevi and Foligno. It was commissioned by consul Gaius Flaminius, renewed under Augustus and had a total length of 336 km.

After the Second Punic War (218–201 BC) and Hannibal's defeat plus the victory over Antiochus III in the battle of Magnesia (190 BC), Rome's hegemony in the Mediterranean region was established, leading to much construction activity. The reparations paid by Carthage financed many projects such as the Via Valeria in Sicily, the Via Aemilia in northern Italy, or the Via Postumia in the Po valley.

During the 2nd c. BC the Via Egnetia is built from Dyrrhachium to Thessaloniki, it is the first road outside Italy. Civil war in Italy hindered construction, although G. Gracchus planned many roads in his *lex Sempronia viaria.* The consul Marius resumes road building and during the time of Sulla and Pompey the first trail crosses Mount Geneva, connecting Rome with upper Italy, Gaul and Spain.

Under Caesar the passage over the Great St. Bernhard is improved, in Spain the Via Augustus is begun. The Via Julia Augusta extends through the Rivièra, following the Mediterranean coast and the prefect Julius Cottius opens a proper road over the Mont Genèvre Pass between upper Italy and eastern Gaul.

Eastwards, the crossings of little St.Bernhard, the Brenner Pass and further roads from Aquileia into Noricum complete the picture.

In scattered places Roman roads with worn paving exist in their original state, but most of them have vanished below modern highways, which invariably follow the Roman alignment. Of the Alpine passes the 2310 m Pass da Sett (Septimerpass) from Bivio to Maloja (Grisons, Switzerland) remains in its original state and is an attraction for mountain bikers.

The greatest number of bridges is in Italy, southern France and Spain. Thriving economic conditions and suitable materials facilitated construction. Of course, for destinations along the Mediterranean coast ships were preferred for transportation. Bridges north of the Alps and in the Balkans were mostly built in timber, which was abundant in this region. In Turkey the enduring Byzantine Empire and Ottoman rule right after, gave some continuity to the maintenance of infrastructure. Few bridges were needed in North Africa since it is lacking rivers.

During the rule of Trajan the Empire reached its greatest extent. In conclusion we can say that the Roman road network was built between 400 BC and 200 AD.

Alphabetical listing of major roads

Via Aemilia: 2nd c. BC. Built in 187 BC it served the Po valley via Arimium, Bononia (Bologna), Placentia (Piacenza), Patuvium (Padua) to Aquilia;

Via Aemilia Scauri: 1st c. BC. From Luni (Luna) via Genoa, Alba Docilia to Dertona;

Via Agrippa: 1st c. AD. In the Rhône valley from Arlete to Lugundunum (Lyon);

Via Appia: 4th c. BC. The principal Roman road from Rome to Brindisi;

Via Augusta : 1st c. AD. Continued the Via Domitia to Gerona, Barcelona, Tarragona, Sagunto, Valencia, Cartagena, Cordoba, Carmona, Hispalis (Sevilla) till Gades (Cadiz);

Via Aurelia: 3rd c. BC. Built about 240 BC from Rome to Pidaae (Pisa) and Luni (Luna);

Via Cassia: 2nd c. BC. From Rome via Arretium and Faesulae (Fiesole) to Pisa;

Via Claudia Valeria: 1st c. AD. Built under Emperor Claudius across the Mons Imeus to Pescar on the Adriatic Sea;

Via Clodia: From Rome to Vetulonia;
Via Collatina: Runs parallel to the Via Tiburtina;
Via Decia: From Teriolis near Innsbruck to Bigantium (Bregenz);
Via Domitia: 2nd c. BC. Connects Italy with Spain, built by Gnaius Domitius Ahenobarbus;
Via Domitiana: 2nd c. BC. From the Via Appia to Terracina and Naples;
Via Egnatia: 2nd c. BC. First road outside Italy, to continue the Via Appia after crossing the sea by boat to Dyrrachium (Durres). Built under Gaius Inarius it went via Edessa and Antipolis till Constantinople;
Via Ermine: From Londonium to York;
Via Flaminia: 3rd c. BC. See previous description above;
Via Gabiniana: In Ex-Yugoslavia;
Via Julia Augusta: 1st c. AD. Along the Mediterranean coast from Genoa via Albintimilium to Arelate (Arles);
Via Latina: 4th c. BC. From Rome past Casinum to Capua;
Via Ostiense: one of the oldest Roman roads from Rome to Ostia;
Via Postumia: 2nd c. BC. Built 148 BC from Geboa via Placentia and Cremona to Verona;
Via Salaria: From Rome via Raete, Asculum to Ancona. The name derives from salt transports;
Via Severiana: 3rd c. AD. From Rome to Terracina;
Via Tiburtina: c. 4th c. BC. From Rome to Tivoli;
Via Traiana: 1st c. AD. An addition to the Via Appia to Beneventum and Brindisi;
Via Valeria: 2nd c. BC. Censor Maximus continued the Via Tiburtina from Tivoli eastward.

Fosse Street: From Isca Dumnoniorium to Lindum (Lincoln);
Stane Street: From Reginorum (Chichester) to London;
Ruta della Plata: In western Spain from Gijon via Asturica Augusta, Salamanca, Caceres, Augusta Emerita to Hispalis (Seville).
Watling Street: From London to Ratae Coritanorum (Leicester).

The Roman road network is depicted in a series of nine maps:

Map 1: **I** Italy (with adjoining areas to the north and of the Balkan)
Map 2: **F** France (with adjoining areas to the east)
Map 3: **SP** Spain and Portugal
Map 4: **A** North Africa from Morocco to Libya
Map 5: **UK** Britain with parts of the continent
Map 6: **D** Danube countries (Germany, Austria, Hungary)
Map 7: **B** Balkan with Greece and the lower Danube countries
Map 8: **T** Turkey
Map 9: **NE** Near East from Turkey to Egypt

All place names are indicated in Latin. The maps carry a reference grid (A–1 etc.). There are two alphabetic lists of place names – one for Latin and one for modern names. They give the location of each place by way of the map initial and the grid figures.

On the following pages a series of nine roadmaps are given, covering the Roman Empire. They are organized according to current political conditions, however all the place names are in Latin. As cross-reference alphabetical listings of the Roman names and modern names are added, together with the plan coordinates of each location. The numbering of bridges follows the same system.

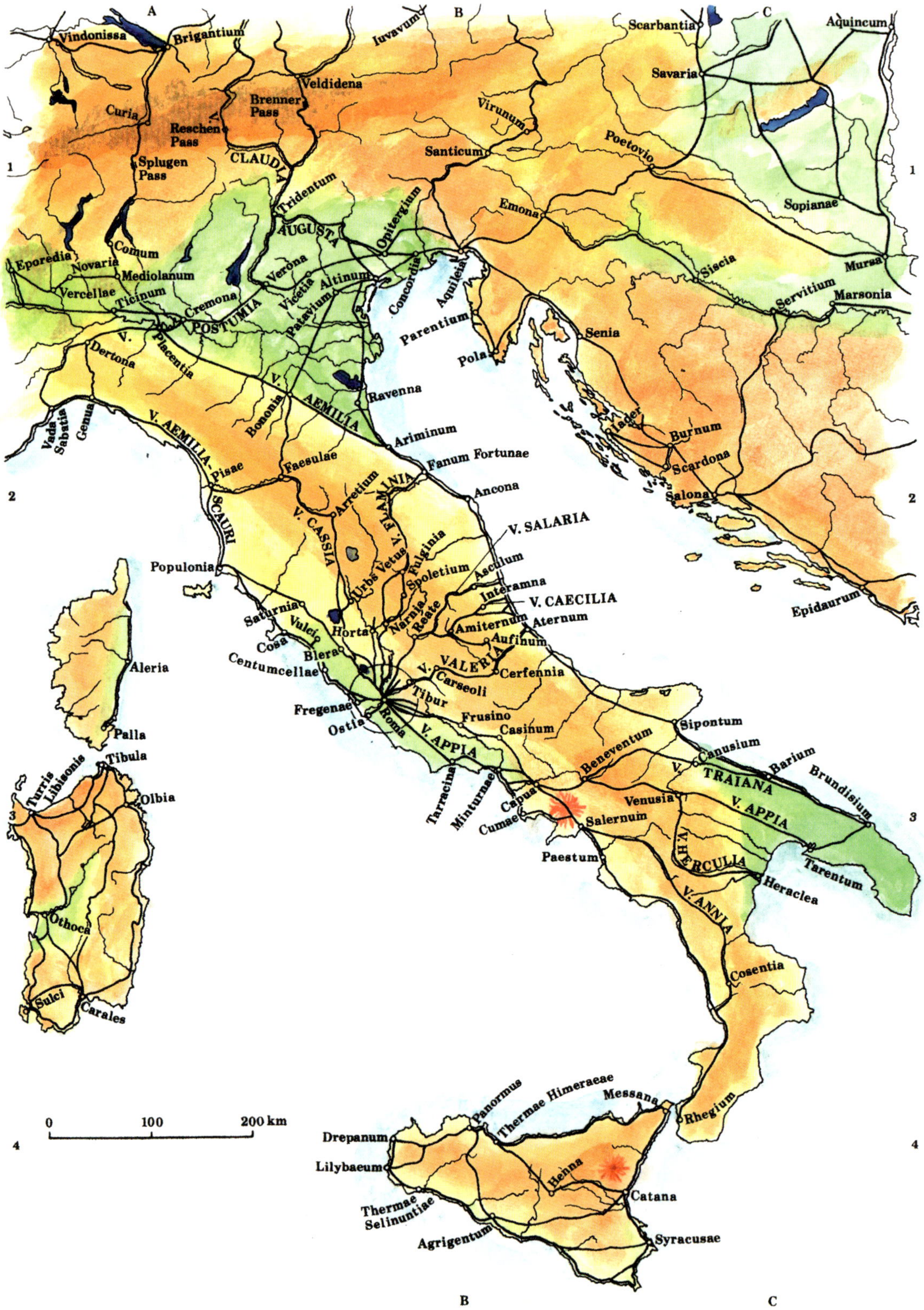
A
B
C
1
2
3
4
Vindonissa
Brigantium
Iuvavum
Scarbantia
Aquincum
Veldidena
Brenner Pass
Curia
Reschen Pass
Splugen Pass
CLAUDIA
Virunum
Savaria
Santicum
Poetovio
Tridentum
Opitergium
Emona
Sopianae
AUGUSTA
Comum
Eporedia
Novaria
Mediolanum
Vercellae
Ticinum
Cremona
Verona
Vicetia
Altinum
Patavium
Concordia
Aquileia
Siscia
Mursa
Servitium
Marsonia
POSTUMIA
V.
Dertona
Placentia
Parentium
Pola
Senia
V.
AEMILIA
Bononia
Ravenna
Vada Sabatia
Genua
V. AEMILIA
Ariminum
Iader
Burnum
Scardona
Salona
Fanum Fortunae
Faesulae
Pisae
SCAURI
Arretium
FLAMINIA
Ancona
V. CASSIA
V. FLAMINIA
V. SALARIA
Populonia
Urbs Vetus
Fulginia
Spoletium
Asculum
Interamna
V. CAECILIA
Epidaurum
Saturnia
Vulci
Cosa
Horta
Narnia
Reate
Amiternum
Aternum
Aufinum
Aleria
Blera
Centumcellae
V. VALERIA
Cerfennia
Carseoli
Tibur
Fregenae
Ostia
Roma
Frusino
Casinum
Sipontum
Palla
Tibula
Turris Libisonis
Olbia
V. APPIA
Beneventum
Canusium
Barium
Brundisium
V. TRAIANA
Tarracina
Minturnae
Capua
Cumae
Salernum
Venusia
V. APPIA
Paestum
V. HERCULIA
Tarentum
Heraclea
V. ANNIA
Othoca
Cosentia
Sulci
Carales
0
100
200 km
Panormus
Thermae Himeraeae
Messana
Rhegium
Drepanum
Lilybaeum
Henna
Catana
Thermae Selinuntiae
Agrigentum
Syracusae

111. Italy (with adjoining areas to the north and of the Balkan).
112. France (with adjoining areas to the east)

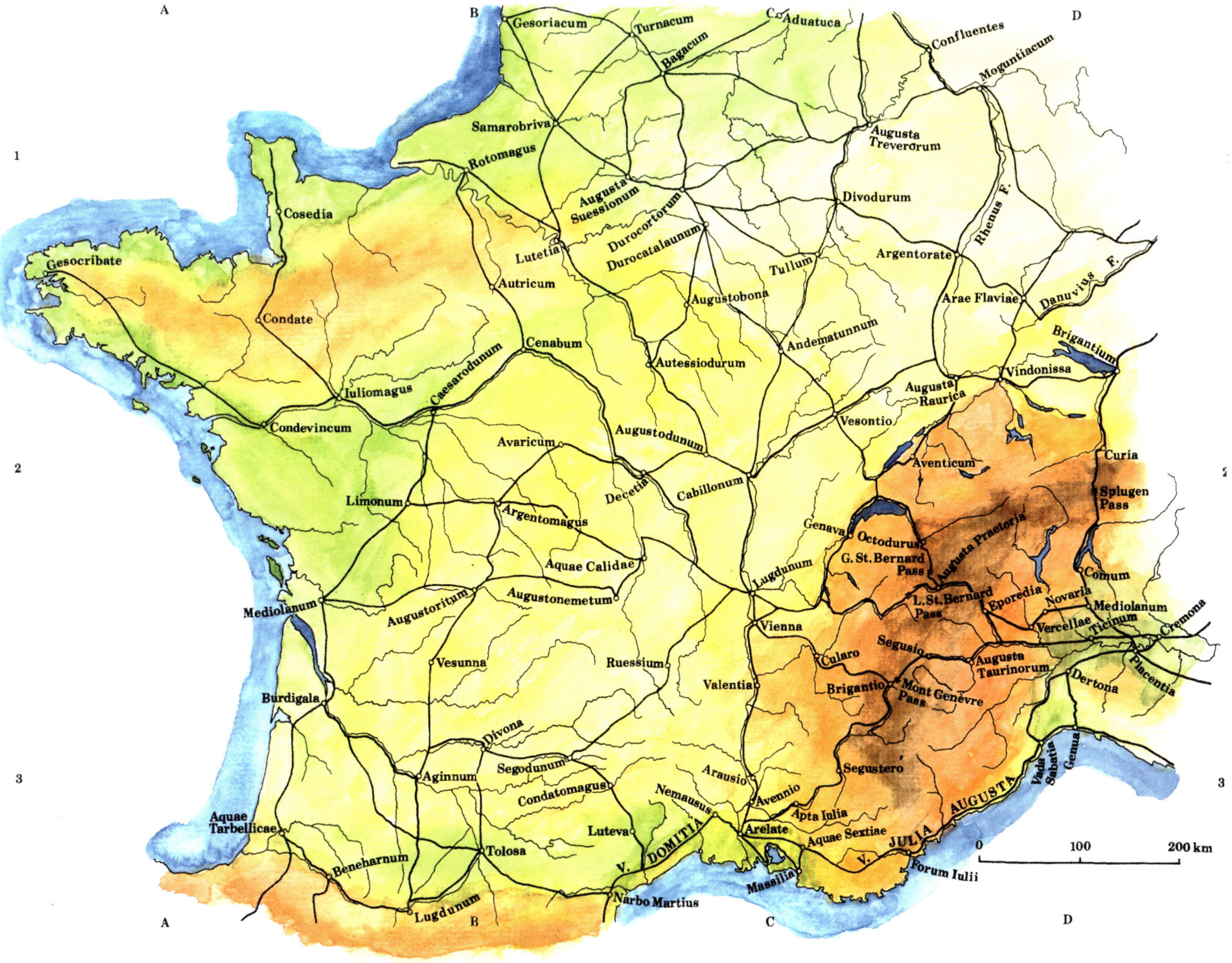

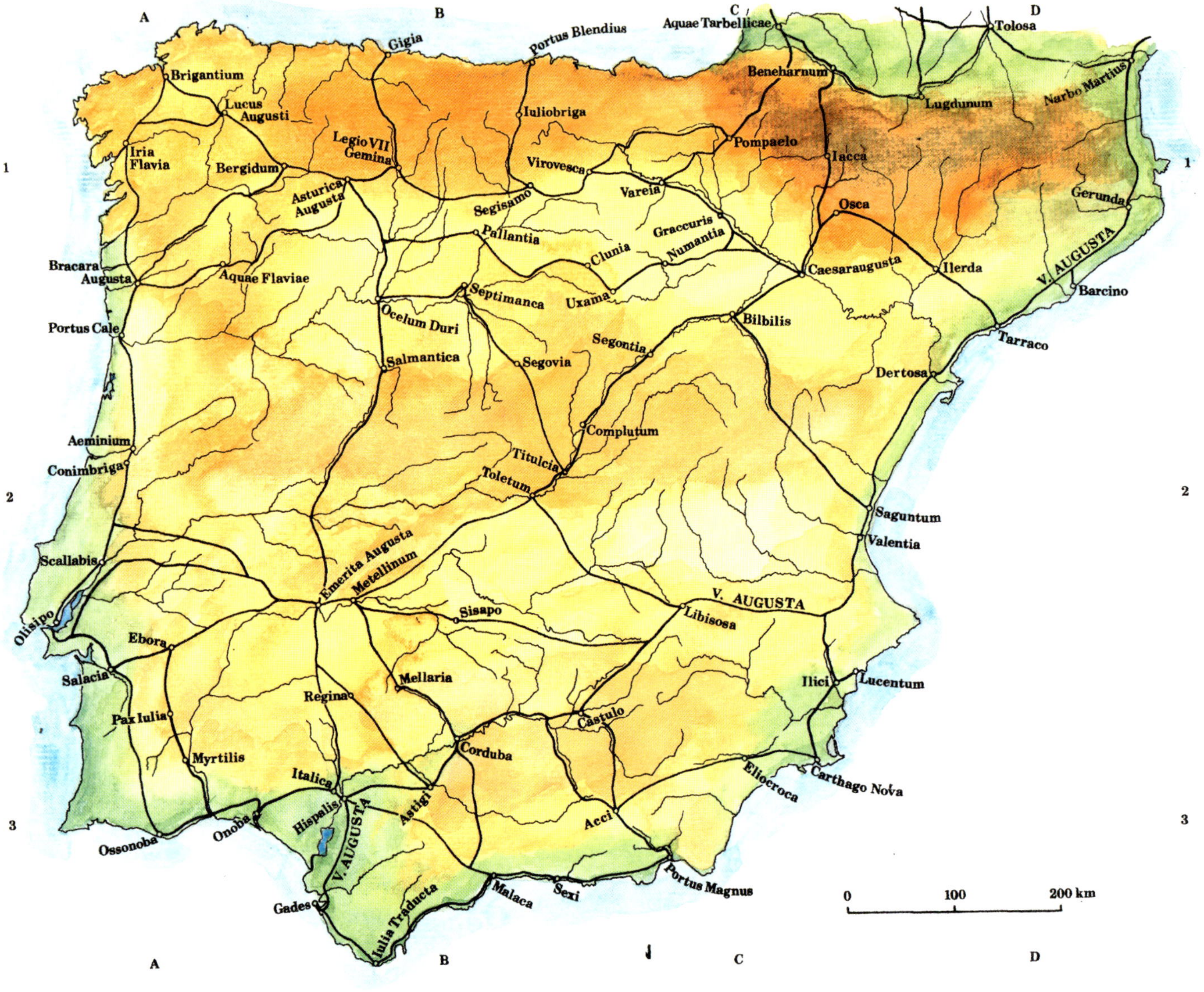
A
B
C
D
1
2
3
Aquae Tarbellicae
Tolosa
Gigia
Portus Blendius
Brigantium
Lucus
Augusti
Iuliobriga
Beneharnum
Lugdunum
Narbo Martius
Legio VII
Gemina
Iria
Flavia
Bergidum
Asturica
Augusta
Virovesca
Pompaelo
Iacca
Segisamo
Vareia
Osca
Gerunda
Pallantia
Graccuris
Numantia
Clunia
Bracara
Augusta
Aquae Flaviae
Caesaraugusta
Ilerda
V. AUGUSTA
Barcino
Septimanca
Uxama
Ocelum Duri
Bilbilis
Portus Cale
Segontia
Tarraco
Salmantica
Segovia
Dertosa
Complutum
Aeminium
Conimbriga
Titulcia
Toletum
Saguntum
Valentia
Scallabis
Emerita Augusta
Metellinum
V. AUGUSTA
Olisipo
Sisapo
Libisosa
Ebora
Salacia
Mellaria
Ilici
Lucentum
Regina
Pax Iulia
Castulo
Corduba
Myrtilis
Italica
Carthago Nova
Eliocroca
Hispalis
Astigi
Acci
Onoba
Ossonoba
V. AUGUSTA
Gades
Iulia Traducta
Malaca
Sexi
Portus Magnus
0
100
200 km

113. Spain and Portugal.
114. North Africa from Morocco to Libya.

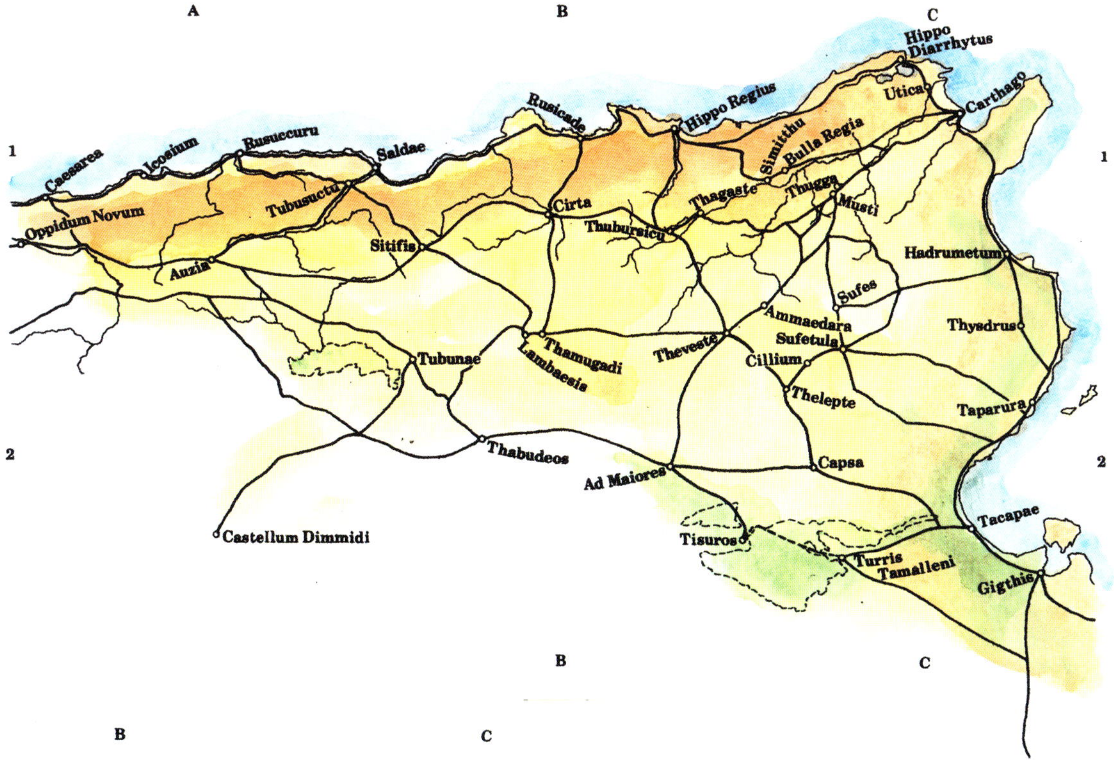

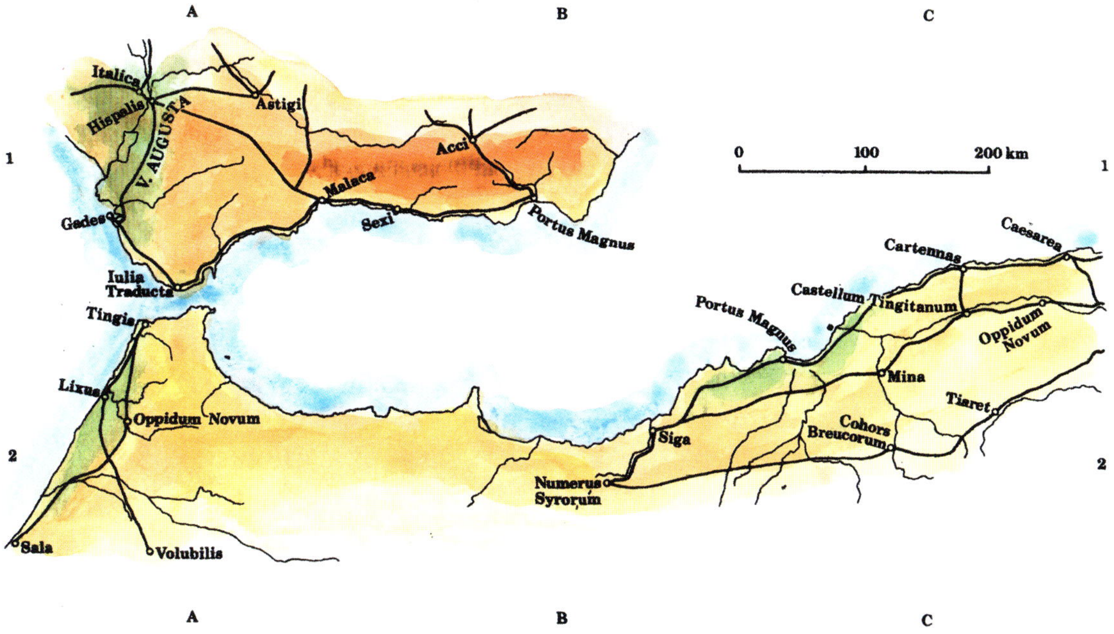

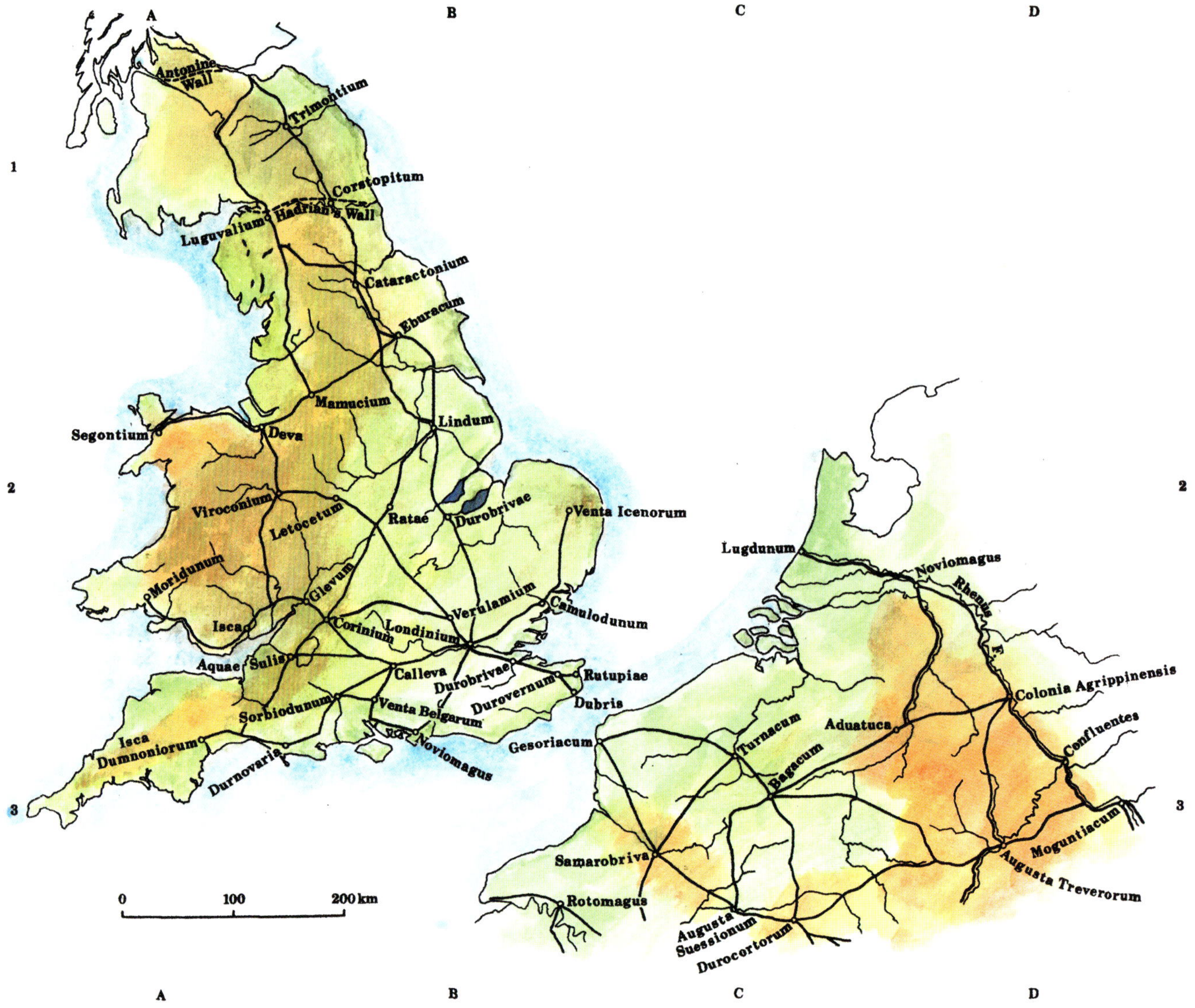
A
B
C
D
1
2
3
Antonine Wall
Trimontium
Corstopitum
Hadrian's Wall
Luguvalium
Cataractonium
Eburacum
Mamucium
Lindum
Segontium
Deva
Viroconium
Letocetum
Ratae
Durobrivae
Venta Icenorum
Moridunum
Glevum
Isca
Corinium
Verulamium
Camulodunum
Londinium
Aquae Sulis
Calleva
Durobrivae
Rutupiae
Durovernum
Dubris
Sorbiodunum
Venta Belgarum
Noviomagus
Isca Dumnoniorum
Durnovaria
Lugdunum
Noviomagus
Rhenus F.
Colonia Agrippinensis
Aduatuca
Confluentes
Gesoriacum
Turnacum
Bagacum
Moguntiacum
Samarobriva
Augusta Treverorum
Rotomagus
Augusta Suessionum
Durocortorum
0
100
200 km
A
B
C
D

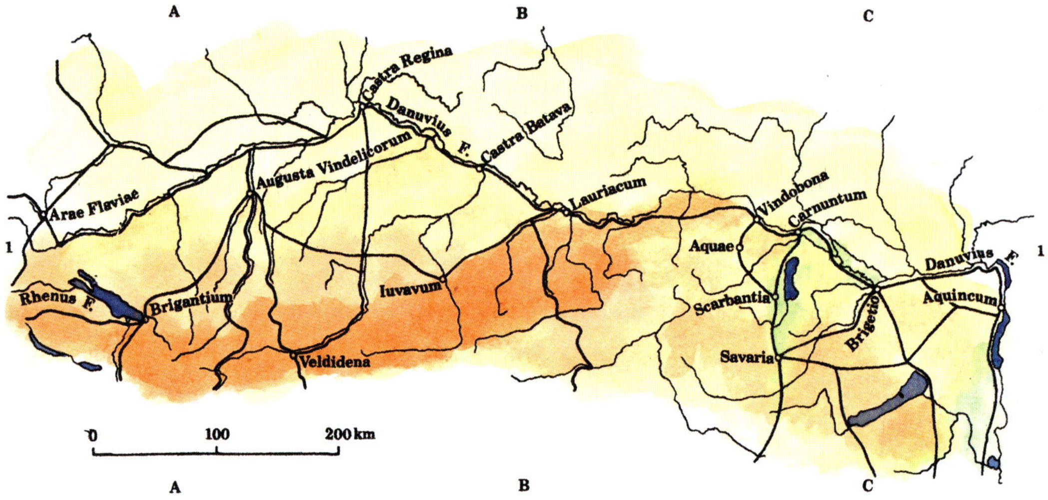
A
B
C
1
1
Castra Regina
Danuvius F.
Castra Batava
Augusta Vindelicorum
Arae Flaviae
Lauriacum
Vindobona
Carnuntum
Aquae
Danuvius F.
Rhenus F.
Brigantium
Iuvavum
Scarbantia
Brigetio
Aquincum
Veldidena
Savaria
0
100
200 km
A
B
C

115. Britain with parts of the continent.
116. Danube countries (Germany, Austria, Hungary).
117. Balkan with Greece and the lower Danube countries.

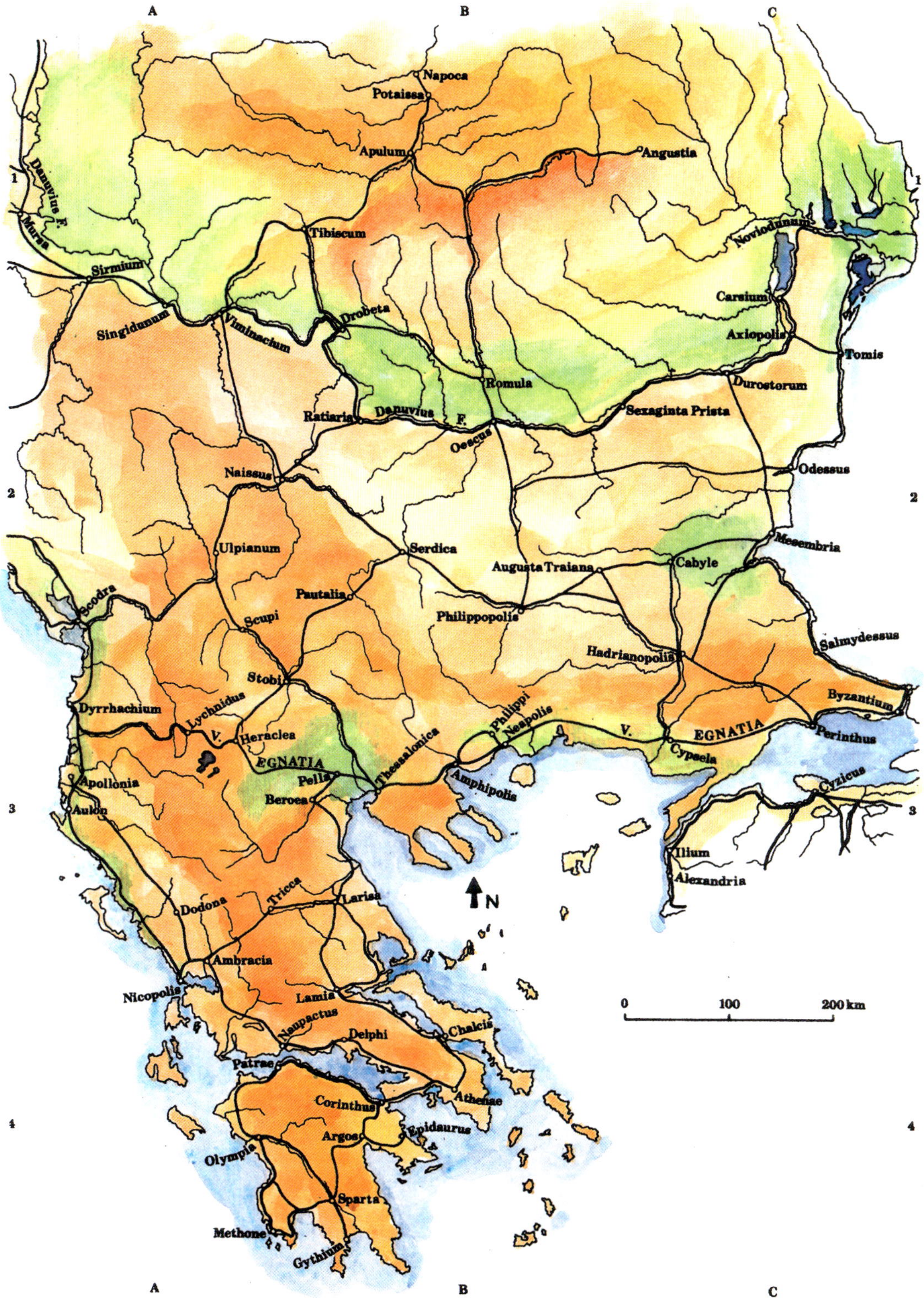

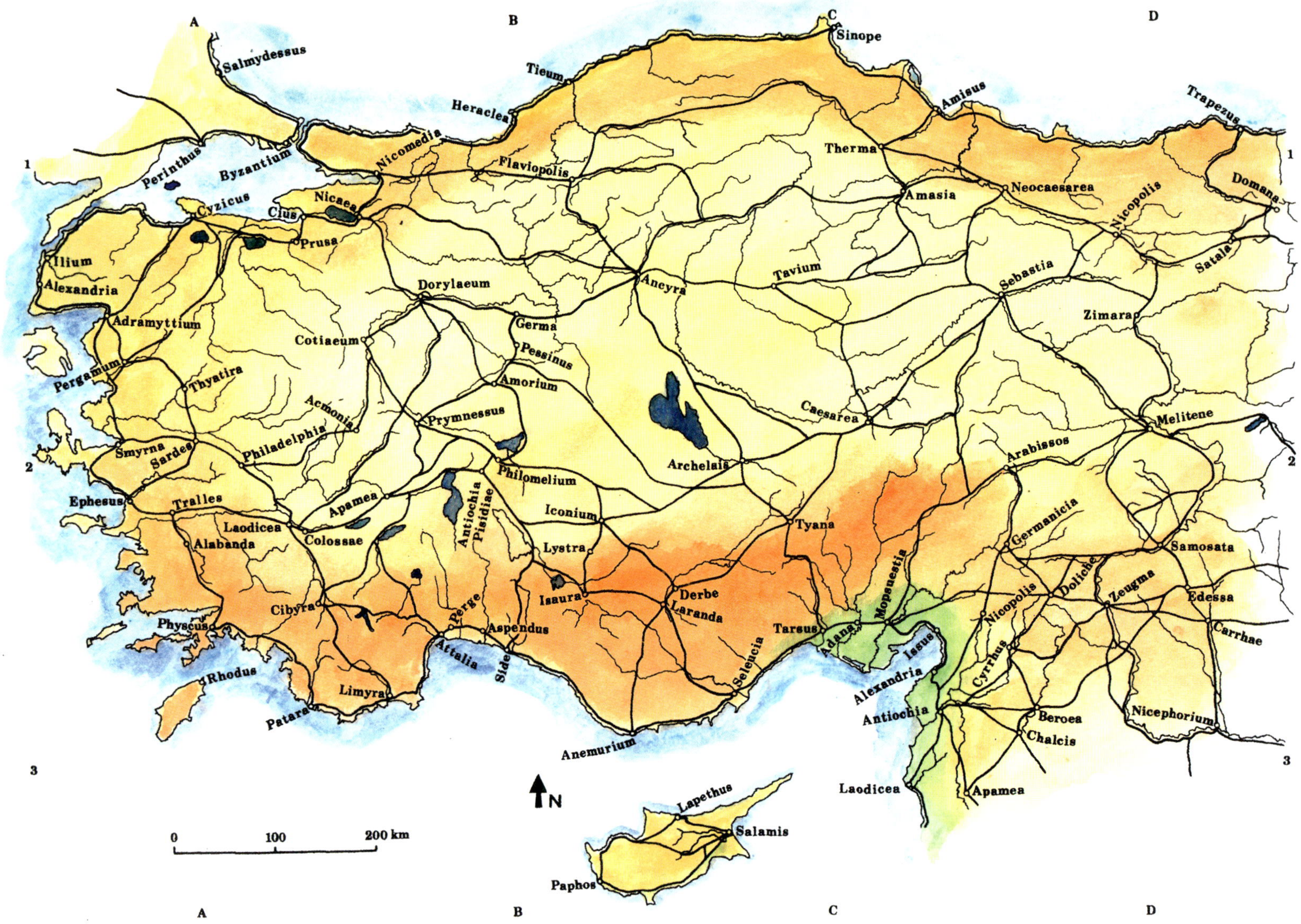

A
B
C
D
Sinope
Salmydessus
Tieum
Heraclea
Amisus
Trapezus
Perinthus
Byzantium
Nicomedia
Therma
Flaviopolis
Amasia
Neocaesarea
Domana
Nicopolis
Cyzicus
Nicaea
Clus
Prusa
Ilium
Alexandria
Satala
Ancyra
Tavium
Sebastia
Dorylaeum
Germa
Adramyttium
Cotiaeum
Pessinus
Zimara
Pergamum
Thyatira
Amorium
Acmonia
Prymnessus
Caesarea
Melitene
Smyrna
Sardes
Philadelphia
Archelais
Arabissos
Philomelium
Ephesus
Tralles
Apamea
Antiochia Pisidiae
Iconium
Tyana
Laodicea
Alabanda
Colossae
Lystra
Germanicia
Samosata
Mopsuestia
Isaura
Derbe
Laranda
Zeugma
Edessa
Cibyra
Perge
Doliche
Nicopolis
Physcus
Aspendus
Tarsus
Adana
Issus
Carrhae
Attalia
Side
Seleucia
Cyrrhus
Rhodus
Limyra
Alexandria
Patara
Antiochia
Beroea
Chalcis
Nicephorium
Anemurium
Laodicea
Apamea
1
2
3
N
0
100
200 km
Lapethus
Salamis
Paphos

118. Turkey.
119. Near East from Turkey to Egypt.

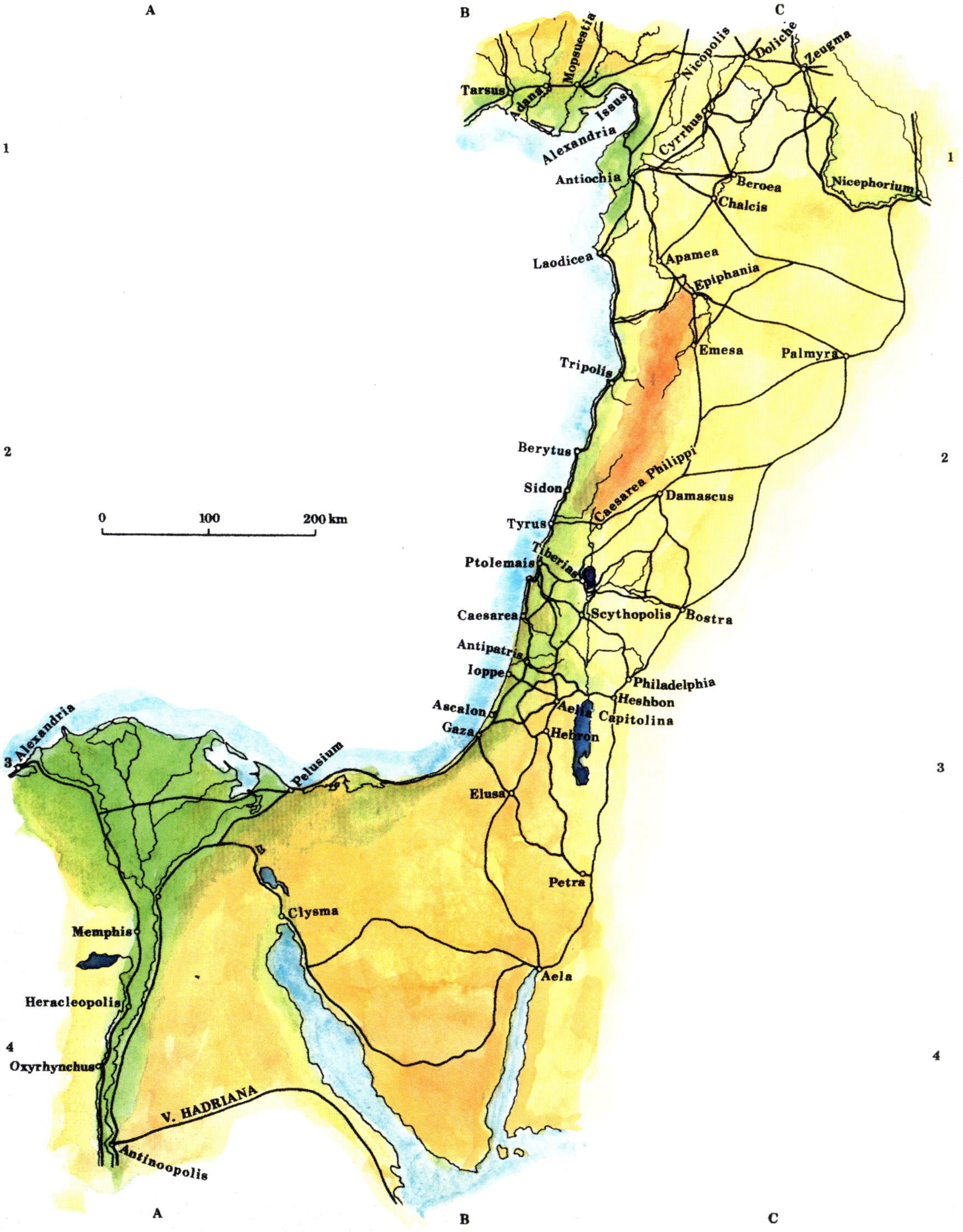

3.2. Listing of place names with plan indication and coordinates

Latin name	Name today	Plan coordinates
Acci	Guadix	SP C-3
Ad Maiores	Henchir Besseriani	A D-2
Adana	Adana	T C-3
Adramyttium	Karatas	T A-1
Aduatuca Tungrorum	Tongeren	F C-1
Aela	Aqaba	NE B-4
Aelia Capitolina	Jerusalem	NE B-3
Aeminium	Coimbra	SP A-2
Aginnum Nitiobrogum	Agen	F B-3
Agrigentum	Agrigento	I B-4
Alabanda	Alabanda	T A-2
Alexandria	Alexandria	T A-1
Alexandria (Egypt)	Alexandria (Egypt)	NE A-3
Alexandria ad Issum	Iskenderun	T C-3
Altinum	Altino	I B-1
Amasia	Amasya	T C-1
Ambracia	Arta	B A-3
Amisus	Samsun	T C–1
Amiternum	San Vittorino	I B-2
Ammaedara	Haidra	A E-1
Amorium	Hergan Kale	T B-2
Amphipolis	Nine Ways	UK B-3
Ancona	Ancona	I B-2
Ancyra	Ankara	T B-1
Andematunnum Lingonum	Langres	F C-2
Anemurium	Anemurium	T B-3
Antinoöpolis	Sheikh Ibada	NE A-4
Antiochia ad Orontem	Antakya	T C-3
Antiochia Pisidae	Yalvac	T B-2
Antipatris	Qal'at Rosha Ayin	NE B-3
Apamea	Dinar	T B-2
Apamea	Qal'at al-Madiq	NE C-1
Apollonia	Apollonia	B A-3
Apta Iulia	Apt	F C-3
Apulum	Alba Iulia	B C-3
Aquae	Calan	D C-1
Aquae Calide	Vichy	F C-2
Aquae Flavie	Chaves	SP A-1
Aquae Sextiae	Aix-en-Provence	F C-3
Aquae Sulis	Bath	UK A-3
Aquae Tarbellicae	Dax	F A-3
Aquileia	Aquileia	I B-1
Aquincum	Budapest	D C-1
Arabissos	Yarpuz	T D-2
Arae Flaviae	Rottweil	D A-1
Arausio	Orange	F C-3
Archelais-Garsaura	Aksaray	T C-2
Arelate	Arles	F C-3
Argentomagus	Argenton	F B-2
Argentorate	Strasbourg	F D-1
Argos	Argos	B B-4
Ariminum	Rimini	I B-2
Arretium	Arezzo	I B-2
Ascalon	Tell Ashqelon	NE B-3
Asculum	Ascoli Piceno	I B-2
Aspendus	Aspendos	T B-3

Latin name	Name today	Plan coordinates
Astigi	Ecija	SP B-3
Asturia Augusta	Astorga	SP B-1
Aternum	Pescara	I B-2
Athenae	Athens	B B-4
Attalia	Antalia	T B-3
Aufinum	Ofena	I B-2
Augusta Praetoria	Aosta	F D-2
Augusta Raurica	Augst	F D-2
Augusta Suessionum	Soisson	F C-1
Augusta Taurinorum	Turin	F D-3
Augusta Traiana	Stara Zagora	B B-2
Augusta Treverorum	Trier	F C-1
Augusta Vindelicorum	Augsburg	D A-1
Augustia	Bretcu	B C-1
Augustobona Tricassium	Troyes	F C-1
Augustodunum Aeduorum	Autun	F C-2
Augustonementum Arvernorum	Clermont-Ferrand	F B-2
Augustoritum Lemovicum	Limoges	F B-2
Aulon	Vlore	A-3
Autessiodurum	Auxerre	F C-2
Autricum Carnutum	Chartres	F B-1
Auzia	Sour-el-Ghozlane	A C-1
Avaricum Biturigum	Bourges	F B-2
Avennio	Avignon	F C-3
Aventicum Helvetiorum	Avenches	F C-2
Axiopolis	Hinog	B C-1
Bagacum Nerviorum	Bavay	F C-1
Barcino	Barcelona	SP D-1
Barium	Bari	I C-3
Beneharnum	Lescar-en-Béarn	F B-3
Beneventum	Benevento	I B-3
Bergidum	Bergidum	SP A-1
Beroea	Aleppo	T D-3
Berytus	Beirut	NE B-2
Bilbilis	Calatayud	SP C-1
Blera	Bieda	I B-2
Bononia	Bologna	I A-2
Bostra	Busra esh-Sham	NE C-2
Bracara Augusta	Braga	SP A-1
Brigantio	Briancon	F C-3
Brigantium	Betanzos	SP A-1
Brigantium	Bregenz	F D-2
Brigetio	Szöny	D C-1
Brundisium	Brindisi	I C-2
Bulla Regia	Bulla Regia	A E-1
Burdigala	Bordeaux	F B-3
Burnum	Ivosevci	I C-2
Byzantium	Istanbul	T A-3
Cabillonum	Chalon-sur-Saône	F C-2
Cabyle	Kabile	B C-2
Caesaraugusta	Zaragoza	SP C-1
Caesarea	Kayseri	T C-2
Caesarea	Qesari	NE B-3
Caesarea Iol	Cherchell	A C-1
Caesarea Philippi	Baniyas	NE B-2
Caesarodunum Turonum	Tours	F B-2
Camulodunum	Colchester	UK B-2
Canusium	Canos di Puglia	I C-3

Latin name	Name today	Plan coordinates
Capsa	Gafsa	A E-2
Capua Vetere	Capua	I B-3
Carales	Cagliari	I A-4
Carnuntum	Petronell	D C-1
Carrhae	Harran	T D-3
Carseoli	Carsoli	I B-3
Carsium	Hirsova	B B-3
Cartennas	Ténès	A C-1
Carthago	Carthage	A E-1
Carthago Nova	Cartagena	SP D-3
Casinum	Cassino	I B-3
Castellum Dimmidi	Messaad	A C-2
Castellum Tingitanum	Ech Cheliff	A C-1
Castra Batava	Passau	D B-1
Castra Regina	Regensburg	D B-1
Castulo	Cazlona	SP B-3
Catana	Catania	I C-4
Cataractonium	Catterick	UK B-1
Cavella Atrebatum	Silchester	UK B-3
Cenabum	Orleans	F B-2
Centumcellae	Civitavecchia	I B-3
Cerfennia	Corfinia	I B-3
Chalcis	Chalkis	B B-4
Chalcis	Qinnisrin	NE C-1
Cibyra	Horzum	T A-2
Cillium	Kasserine	A E-2
Cirta	Constantine	A D-1
Cius	Gemlik	T A-1
Clunia	Coruna del Conde	SP B-1
Clysma	Suez	NE A-4
Cohors Breucorum	Takhemaret	A C-2
Colonia Agrippinensis	Cologne	UK D-3
Colossae	Colossae	T A-2
Complutum	Alcala de Henares	SP B-2
Comum	Como	I A-1
Concordia	Concordia	I B-1
Condate Redonum	Rennes	F A-2
Condatomagus	Millau	F B-3
Condevincum Namnetum	Nantes	F A-2
Confluentes	Koblenz	F C-1
Conimbriga	Condeixa-a-Velha	P A-2
Corduba	Córdoba	SP B-3
Corinium Dobunnorum	Cirencester	UK B-2
Corinthus	Corinth	B B-4
Corstopidum	Corbridge	UK B-1
Cosa	Cosa	I A-2
Cosedia	Coutances	F A-1
Cosencia	CosenzaI	C-3
Cotiaeum	Kütahya	T B-2
Cremona	Cremona	I A-1
Cularo	Grenoble	F C-3
Cumae	Cuma	I B-3
Curia	Chur	I A-1
Cypsela	Ipsala	B C-3
Cyrrhus	Horos	T D-1
Cyzicus	Balkis Kale	T A-3
Damascus	Damascus	NE C-2
Danubius Fluvius	Danube River	D

Latin name	Name today	Plan coordinates
Decetia	Decize	F C-2
Dertona	Tortona	I A-2
Dertosa	Tortosa	SP D-2
Deva	Chester	UK A-2
Divodurum Mediomatricorum	Metz	F C-1
Divona Cadurcorum	Cahors	F B-3
Doliche	Tell Duluk	T D-2
Domana	Tomna	T D-1
Dorylaeum	Sar Hüyuk	T B-1
Drepanum	Trapani	I B-4
Drobeta	Turnu-Severin	B B-1
Dubris	Dover	UK B-3
Durnovaria	Dorchester	UK A-3
Durobrivae	Rochester	UK B-3
Durobrivae	Water Newton	UK B-2
Durocortorum Remorum	Reims	F C-1
Durostorum	Silistra	B C-2
Durovernum Cantiacorum	Canterbury	UK B-3
Dyrrhachium	Durres	B-3
Ebora	Evora	SP A-2
Eburacum	York	UK B-2
Edessa	Urfa	T D-2
Eliocroca	Lorca	SP D-3
Elusa	Haluza	NE B-3
Emerita Augusta	Mérida	SP B-2
Emesa	Homs	NE C-2
Emona	Ljubljana I	B-1
Ephesus	Selcuk	T A-2
Epidaurum	Cavtat	I C-2
Epiphania	Hama	NE C-1
Eporedia	Ivrea	I A-1
Faesulae	Fiesole	I A-2
Fanum Fortunae	Fano	I B-2
Flaviopolis-Crateia	Gerede	T B-1
Fregenae	Fregene	I B-3
Frusino	Frosinone	I B-3
Fulginia	Foligno	I B-2
Gades	Cadiz	SP B-3
Genava	Geneva	F C-2
Genua	Genoa	I A-2
Germa	Karaca Pasa Ören	T B-1
Germanicia	Maras	T D-2
Gerunda	Gerona	SP D-1
Gesocribate	Brest	F A-1
Gesoriacum	Boulogne	F B-1
Gigia	Gijon	SP B-1
Gigthis	Gighti	A E-2
Glevum	Gloucester	UK A-2
Graccuris	Alfaro	SP C-1
Gythium	Gytheion	B B-4
Hadrianopolis	Edirne	B C-2
Hadrumetum	Sousse	A E-1
Henna	Enna	I B-4
Heraclea	Policoro	I C-3
Heraclea Pelagonia	Bitola	B A-3
Heraclea Pontica	Eregli	T B-1
Heracleopolis	Ihnasya el-Madina	NE A-4
Heshbon	Hisban	NE B-3

Latin name	Name today	Plan coordinates
Hippo Diarrhytus	Bizerta	A E-1
Hippo Regius	Annaba	A D-1
Hispalis	Seville	SP B-3
Horta	Orte	I B-2
Iacca	Jaca	SP C-1
Iader	Zadar	I B-2
Iconium	Konya	T B-2
Icosium	Algier	A C-1
Ilerda	Lerida	SP D-1
Ilici	Elche	SP D-3
Ilium	Troy	T A-1
Interamna	Teramo	I B-2
Ioppe	Jaffa	NE B-3
Iria Flavia	Padron	SP A-1
Isaura	Zengibar Kale	T B-2
Isca	Caerleon	UK A-2
Isca Dumnoniorum	Exeter	UK A-3
Italica	Santiponce	SP B-3
Iulia Traducda	Tarifa	SP B-3
Iuliobriga	Retortilla	SP B-1
Iuliomagnus Andecavorum	Angers	F B-2
Iuvavum	Salzburg	D B-1
Lambaesis	Lambese	A D-2
Lamia	Lamia	A D-2
Laodicea	Eski Hisar	T A-2
Laodicea ad Mare	Latakia	NE C-1
Laranda	Karaman	T C-2
Larisa	Larisa	B B-3
Lauriacum	Lorch	D B-1
Legio VII Gemina	León	SP B-1
Letocetum	Wall	UK B-2
Libisosa	Lezuza	SP C-2
Lilibaeum	Marsala	I B-4
Limonum Pictonum	Poitiers	F B-2
Limyra	Limyra	T B-3
Lindum	Lincoln	UK B-2
Lixus	Chemmich	A A-2
Londinium	London	UK B-2
Lucentum	Alicante	SP D-3
Lucus Augusti	Lugo	SP A-1
Lugdunum	Lyon	F C-2
Lugdunum Batavorum	Katwijk	UK C-2
Lugdunum Convenarum	St Bertrand de Comminges	F B-3
Luguvalium	Carlisle	UK A-1
Lutetia Parisiorum	Paris	F B-1
Luteva	Lodève	F C-3
Lychnidus	Ohrid	B A-3
Lystra	Hatunsaray	T B-2
Malaca	Malaga	SP B-3
Mamucium	Manchester	UK B-2
Marsonia	Slavonska Brod	I C-1
Massilia	Marseille	F C-3
Mediolanum	Milan	I A-1
Mediolanum Santonum	Saintes	F B-2
Melitene	Malatya	T D-2
Mellaria	Fuente Ovejuna	SP B-3
Memphis	Mit Rahina	NE A-4
Mesembria	Nesebur	B C-2

Latin name	Name today	Plan coordinates
Messana	Messina	I C-4
Metellinum	Medellin	SP B-2
Methone	Methone	B A-4
Mina	Relizane	A C-2
Minturnae	Minturnol	B-3
Moguntiacum	Mainz	F D-1
Moridunum	Carmathen	UK A-2
Mursa	Osijek	B A-1
Murtuna	Perge	T B-3
Musti	Musti	A E-1
Myrtilis	Mertola	SP A-3
Napoca	Cluj	B B-1
Narbo Martius	Narbonne	F B-3
Narnia	Narni	I B-2
Naupactus	Naupactos	B A-4
Neapolis	Kavala	B C-3
Nemausus	Nîmes	F C-3
Neocaesareia-Cabira	Niksar	T D-1
Nicaea	Iznik	T B-1
Nicephorium	Ar-Raqqah	NE C-2
Nicomedia	Izmit	T B-1
Nicopolis	Islahiye	T D-2
Nicopolis	Purk	T D-1
Noviodunum	Isaccea	B C-1
Noviomagus Batavorum	Nijmegen	UK D-2
Noviomagus Regnorum	Chichester	UK B-3
Numantia	Soria	SP C-1
Numerus Syrorum	Maghnia	A B-2
Ocelum Duri	Zamora	SP B-1
Odessus	Varna	B C-2
Oescus	Oescus	B B-2
Olbia	Olbia	I A-3
Olisipo	Lisbon	SP A-2
Olympia	Olympia	B A-4
Onoba	Huelva	SP C-1
Opitergium	Oderzo	I B-1
Oppidum Novum	Ain Defla	A C-1
Oppidum Novum Tingitana	Ksar-el-Kwebir	A A-2
Osca	Huesca	SP C-1
Ossonoba	Faro	SP A-3
Ostia Antica	Ostia	I B-3
Othoca	Oristano	I A-3
Oxyrhynchus	El Bahnasa	NE A-44
Paestum	Pesto	I B-3
Palla	Porto-Vecchio	I A-3
Pallantia	Palencia	SP B-1
Palmyra	Tadmor	NE C-2
Panormus	Palermo	I B-4
Paphos	Paphos	T B-3
Parentium	Porec	I B-1
Patara	Kelemis	T A-3
Patavium	Padua	I B-1
Patrae	Patrai	B A-4
Pautalia	Kjustendil	B B-2
Pax Julia	Beja	SP A-3
Pella	Pella	SP B-3
Pelusium	Tell el-Farama	NE A-3
Pergamum	Bergama	T A-2

Latin name	Name today	Plan coordinates
Perge	Murtuna	T B-3
Perinthus	Marmarae Reglisi	B C-3
Pessinus	Balahisar	T B-2
Petra	Petra	NE B-3
Philadelphia	Alashehir	T A-2
Philadelphia	Amman	NE B-3
Philippi	Philippi	B B-3
Philippopolis	Plovdiv	B B-2
Philomelium	Akshehir	T B-2
Physcus	Marmaris	T A-3
Pisae	Pisa	I A-2
Placentia	Piacenza	I A-1
Poetovio	Ptuj	I C-1
Pola	Pula	I B-2
Pompaelo	Pamplona	SP C-1
Populonia	Populonia	I A-2
Portus Blendius	Suances	SP B-1
Portus Cale	Oporto	SP A-2
Portus Magnus	Almeria	SP C-3
Portus Magnus	Arzew Vieil	A B-2
Potaissa	Turda	B B-1
Praeneste	Palestrina	I
Prusa	Bursa	T A-1
Prymnessus	Afyon	T B-2
Ptolemais	Acre	NE B-2
Ratae Coritanorum	Leicester	UK B-2
Ratiaria	Archar	B B-2
Ravenna	Ravenna	I B-2
Reate	Rieti	I B-2
Regina	Casas de Reina	SP B-3
Rhegium	Reggio de Calabria	I C-4
Rhenus Fluvius	Rhine River	F C-1–D-1
Rhodus	Rhodes	T A-3
Roma	Rome	I B-3
Romula	Resca	B B-2
Rotomagus Veliocassium	Rouen	F B-1
Ruessium Velaviorum	St Paulien en Velay	F C-3
Rusicade	Skikda	A D-1
Rusuccuru	Dellys	A C-1
Rutupiae	Richborough	UK B-3
Sala	Chella	A A-2
Salacia	Alcacer do Sol	SP A-3
Saldae	Bejaia	A C-1
Salernum	Salerno	I B-3
Salmantica	Salamanca	SP B-2
Salmydessus	Midye	B C-2
Salona	Solin	I C-2
Samarobriva	Amiens	F B-1
Samosata	Samsat	T D-2
Santicum	Villach	I B-1
Sardes	Sardes	T A-2
Satala	Sadag	T C-2
Savaria	Szombathely	D C-1
Scallabis	Santarem	SP A-2
Scarbantia	Sopron	D C-1
Scardona	Skradin	I C-2
Scodra	Shkoder	B A-2
Scupi	Skopje	B A-2

Latin name	Name today	Plan coordinates
Scythopolis	Tell Bet She'an	NE B-3
Sebastia-Megalopolis	Sivas	T D-1
Segisamo	Sasamon	SP B-1
Segodunum Rutenorum	Rodez	F B-3
Segontia	Sigüenza	SP C-2
Segontium	Caernarvon	UK C-2
Segustero	Sisteron	F C-3
Seleucia ad Calycadnum	Silifke	T C-3
Senia	Senj	I B-2
Septimanca	Simancas	SP B-1
Serdica	Sofia	B B-2
Servitium	Bosanska Gradisca	I C-1
Sexaginta Prista	Ruse	B C-2
Sexi	Almunécar	SP B-3
Side	Side	T B-3
Sidon	Saida	NE B-2
Siga	Siga	A B-2
Simitthu	Chemtou	A E-1
Singidunum	Belgrade	B A-1
Sinope	Sinop	T C-1
Sipontum	Siponto	I C-3
Sirmium	Sremska-Mitrovica	B A-1
Sisapo	Almaden	SP B-2
Siscia	Sisak	I C-1
Sitifis	Sétif	A D-1
Siulci	Sant' Antonio	I A-4
Smyrna	Izmir	T A-2
Sopianae	Pécs	I C-1
Sorbiodunum	Old Sarum	UK B-3
Sparta	Sparta	B B-4
Spoletium	Spoleto	I B-2
Stobi	Stobi	B A-3
Sufes	Sufes	A E-1
Sufetula	Sbeitla	A E-2
Sulci	Sant' Antioco	I A-4
Syracusae	Siracusa	I C-4
Tacapae	Gabès	A E-2
Taparura	Sfax	A E-2
Tarentum	Taranto	I C-3
Tarracina	Terracina	I B-3
Tarraco	Tarragona	SP D-2
Tarsus	Tarsus	T C-3
Tavium	Büyük Nefezköy	T C-3
Thabudeos	Thouda	A D-2
Thagaste	Souk Ahras	A D-1
Thamugadi	Timgad	A D-2
Thelepte	Thelepte	A E-2
Therma	Merzifon	T C-1
Thermae Himeraeae	Termini Imerese	I B-4
Thermae Selinuntiae	Sciacca	I B-4
Thessalonica	Thessaloniki	B B-3
Theveste	Tébessa	A D-2
Thubursicu Numidarum	Khémissa	A D-2
Thugga	Dougga	A E-1
Thyatira	Akhisar	T A-2
Thysdrus	El Djem	A E-2
Tiaret	Tiaret	A C-2
Tiberias	Tiberias	NE B-2

Latin name	Name today	Plan coordinates
Tibiscum	Jupa	B A-1
Tibula	Capo Testa	I A-3
Tibur	Tivoli	I B-3
Ticinum	Pavia	I A-1
Tieum	Hisärönu	T B-1
Tingis	Tangier	A A-2
Tisuros	Tozeur	A E-2
Titulcia	Titulcia	SP B-2
Toletum	Toledo	SP B-2
Tolosa	Toulouse	F B-3
Tomis	Constanta	B C-2
Trapezus	Trabzon	T D-1
Tricca	Trikala	B A-3
Tridentum	Trento	I A-1
Trimontium	Newstead	UK A-1
Tripolis	Trablus	NE B-2
Tubunae	Tobna	A C-1
Tubusuctu	Tiklat	A C-1
Tullum Leucorum	Toul	F C-1
Turnacum	Tournai	F C-1
Turris Libisonis	Porto Torres	I A-3
Turris Tamalleni	Telmine	A E-2
Tyana	Kemer Hisar	T C-2
Tyrus	Tyre	NE B-2
Ulpianum	Gracanica	B A-2
Urbs Vetus	Orvieto	I B-2
Utica	Utica	A E-1
Uxama	Burgo de Osma	SP C-1
Vada Sabatia	Vado	I A-2
Valencia	Valencia	SP C-2
Valentia	Valence	F C-3
Vareia	Logrono	SP C-1
Veii	Veio	I B-3
Veldidena	Wilten	D A-1
Venta Belgarum	Winchester	UK B-3
Venta Icenorum	Caistor	UK B-2
Venusia	Venosa	I C-3
Vercellae	Vercelli	I A-1
Verona	Verona	I A-1
Verulanium	St Albans	UK B-2
Vesontio	Besancon	F C-2
Vesunna Petrocoriorum	Périgueux	F B-3
Vicentia	Vicenza	I A-1
Vienna	Vienne	F C-2
Viminacium	Kostolac	B A-1
Vindobona	Vienna	D C-1
Vindonissa	Windisch	F D-2
Viroconium	Wroxeter	UK A-2
Virovesca	Briviesca	SP B-1
Virunum	Zollfeld	I B-1
Volubilis	Volubilis	A A-2
Vulci	Vulci	I B-2
Zeugma	Balkis	T D-2
Zimara	Zimarra	T D-2

3.3. Historical maps of the Roman road network

When planning a trip the Romans had various *itineraria* for reference. An *itinerarium danotatum* contained information about distances, road condition, accommodation and toll stations. Similar to modern travel guides the itineraries had pictograms to facilitate their use. The »stone of Tongres« depicts the road from Bonn to Worms in Germany, the »stone of Autun« describes the way from Rome to Augustodunum.

The two best known manuscripts are the *Itinerarium provinciarum Antonini Augusti* from the time of Diocletian (AD 284–305), which is based on a model from the time of Caracalla (AD 211 to 217), and the *Itinerarium Burdigalense*, the Bordeaux itinerary, also known as the *Itinerarium Hierosolymitanum*, or Jerusalem inventory. It was written by an anonymous pilgrim from Burdigala, present-day Bordeaux. It tells of the writer's journey to the Holy Land in 333/334, by land through northern Italy and the Danube valley to Constantinople, then through Asia Minor to Jerusalem. He returned by way of Macedonia, Otranto, Rome, and Milan.

Pliny the Elder and Strabo, both AD 1st c., prepared travel itineraries giving very exact distances. Further sources are a treatise of geography by Ptolemy (AD 127–151) and a cosmography by Anonymous of Ravenna.

A unique record of the road connections as known by the Romans in the fourth century AD is the *Tabula Peutingeriana*, a road map (*Itinerarium pictum*) which covers a stretch of over 100,000 km from the Atlantic to the Gulf of Bengal. It consists of eleven parchment segments, in all 6.82 m long and 34 cm wide. It gives a rather distorted picture, north-south distances are shrunk and west–east distances exaggerated. The Mediterranean is reduced to a narrow strip; north orientation is adjusted to fit the drawing sheets. It was copied by a monk of the 13th c.

The council scribe Konrad Peutinger in Augsburg obtained it in 1508 from the humanist Konrad Celtus as the *Itinerarium Antonini*, since then it is known as *Tabula Peutingeriana.* Eventually, the famous Prince Eugène of Savoy bought the document. He bequeathed it to Emperor Charles VI who put it into the imperial library where it is still kept in Vienna.

The map shows crossings, bridges, passes, rest stops, towns and distances. The three most important cities Rome, Constantinople and Antioch display their patrons. There are symbols for altars, temples, baths, grain stores, settlements, harbors, lighthouses, passes and tunnels.

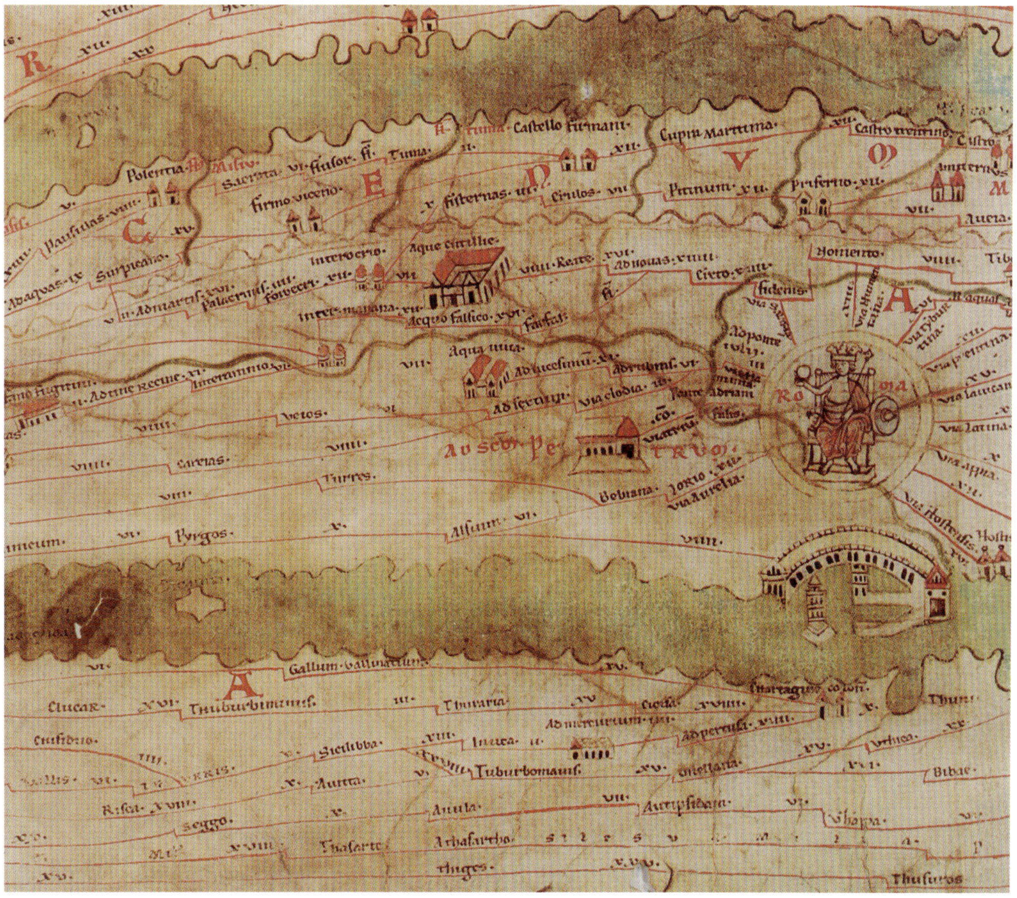

120. Part of segment IV of the *Tabula Peutingeriana*. Italy is in the middle with Rome to the right.

3.4. Excursus: Travelling on Roman roads

Although the roads were built generally for strategic reasons, they soon became part of the empire's wide infrastructure, secure and benefitting everybody. Merchants with their goods constituted the bulk of the traffic, travelling at their own risk and cost. They transported agricultural and industrial products, at this stage mostly of artisanal origin.

There were also private travelers, going on vacation, to a spa or to a festival, as well as pilgrims, to visit one of the many sacred places. To know travelling time, sundials were used. Vitruvius (X 9) even describes a device to record distances. On a wheel of known circumference a cog was attached, which activated with every turn a counting mechanism, which in turn released after every mile a stone into a bucket. Counting the stones one knew the distance travelled.

The most common type of vehicle was known as *carpentum* (hence »car«). If an axle was broken or a wheel was damaged, one could call for a mechanic or *carpentarius* (hence »carpenter«) to repair it. The poor construction of carts and the stone paving made for rather uncomfortable travelling, and it is astonishing, how Cicero or Seneca could read and write while on the road. Most people were walking or rode on a donkey or mule. Oxen served mostly as draft animals.

The roads were equipped with taverns (*tabernae*), inns (*mansiones*) for food and accommodation and stations to change mounts or draft animals (*mutationes*). There were *forae* for the administration and maintenance of the roads, largely funded by the population served. The good condition of the roads translated into fast traffic. The legions were trained to cover 30 km per day. On a trip from Beneventum to Taranto, Cicero attained a daily average of 48 km. Caesar was known for his speed and achieved from Rome to Geneva (Switzerland) a daily average of 150 km.

Augustus established the *cursus publicus,* a postal system also serving for journeys of officials. To use the postal service, travelling officials had to obtain a sort of passport (*diplomata* or *evectio*), issued by the imperial office. The principal users of the *cursus publicus* were imperial couriers, who not only dispatched messages at 150 km per day, but also had to report about conditions in the empire. Known as *speculators* or *agentes in res*, they were not popular.

Military operations sometimes brought traffic to a halt. At certain times and places traffic was so heavy, as to create traffic jams and overcrowding of accommodations. Aurelius Victor, imperial biographer (*liber de Caesaribus* 13) sums up: »To quickly send messages from all parts of the empire, public stations were introduced. They are a very practical institution, but the greed and tyranny of their administrators contributes to the decline of the empire. There is no good or bad public service, which could not be turned into its opposite by the character of the man in charge.«

The Roman poet Horace gives a vivid account of the joys and horrors during a voyage on the Via Appia from Rome to Brindisi in the year 37 BC (Satires 1.5). Accompanied by Maecenas it took them 14 days to cover the 555 km. Their complaints are many: *propter aquam, quod erat detterima, ventri indico bellum* (stomach aches because of foul water), *mali culices ranaeque palustres avertunt samnos* (biting mosquitoes and croaking frogs disturb their sleep), *oculis ... nigra meis colluyria lippus inlinere* (their eyes are hurting and are lined with black ointment), *villa recipisset lacrimoso non sine fumo* (an inn full of caustic smoke from wet firewood), and last but not least *stultissimus usque puellamad mediam noctem expecto* (a failed amorous tête à tête).

4. Technical details of road construction

A eulogy by Statius (silvae, IV 3) describes the construction procedure of the Via Domitiana:

First of all ditches have to be prepared to mark the edges, then we remove the earth as deep as necessary.

Then, we fill the trench with appropriate material, to form the basis of the road bed, which should not settle, as a weak foundation is treacherous for the paving.

Then we cover the road with gravel, tightly packed with wedging stones.

Many hands unite in this work, some fell timber, others clear the ground, some are smoothing the rocks and the edges of the road with iron tools.

Then some lay the paving in hot sand and finely grained tufa.

Some are draining ponds with care. They may refill again, so others lead the streamlets' water away.

Usually roads began as ill defined tracks, they were consolidated by spreading gravel or debris. The decisive improvement was the use of lime mortar. Mixed with aggregate at the ratio 1 to 3, a sort of concrete was produced. From 150 BC onward, lime was replaced by puzzolana

earth, discovered by the Romans around Mt. Vesuvius and later mined in all volcanic areas of the empire. It has the great advantage to solidify without oxygen, that is below water or within bulk masonry, resulting in a waterproof concrete, like the product of today. It arguably was the most important Roman discovery in building technology, giving unprecedented strength to their structures, as well as speeding up the work.

Vitruvius (VII, 1,1) describes the preparation of the floor of a building (*pavimenta*). It is similar to the construction of a road, consisting of a foundation (base course, *statumen*) made of fist-size stones. The next layer called *rudus* is made of broken brick or gravel, bonded with lime mortar. After this follows the so-called *nucleus* of smashed earthware or brick in lime mortar, to receive the actual finish of tiles or mosaic. Vitruvius does not address road construction, but archaeological studies give us the necessary information.

The typical road section is somewhat heavier than the floor described by Vitruvius. It consists of four layers, two as substructure, spreading the loads and protecting from frost, and two finishing layers. First, the ground is well compacted with rammers and rollers and smoothed with a topping of sand. Then squared stones are placed to form the approximately 40 cm thick *statumen*. It extends beyond the actual width of the road and defines the total width of the highway. The edges of the roadway are secured by side walls, to delimit the finishing courses. These consist of a 30 to 50 cm thick rubble base (*rudus*) and a foot thick concrete core (*nucleus*), which receives the paving (*pavimentum*) of stone slabs, or at least a layer of gravel. *Rudus* and *nucleus* were not premixed and cast in place, but rather built up like masonry in mortar.

The pavers reached 1 m^2 in size. 10–15 cm thick, they could weigh 300 kg and more. The pavers were made of local material such as basalt, granite or travertine. Around Rome peperino, a conglomerate of gravel and volcanic ash was frequently used.

In his discussion of the struggles with the Goths, Prokop (500–560 AD) describes the paving of the Via Appia:

»Appius brought the paving from far away; it had the size of millstones and was very hard. There is no such stone in our area. Appius had the slabs squared and smoothed. They were set so tightly that no mortar was needed. The stones are so well set that the observer believes the paving is of one piece, without joints. And though day by day many carriages drive over it and many animals trample it, no stone has become loose or is broken« (Plat. V 14)

The paving had a camber to shed storm water. On either side of the road about 3 m wide trenches (*sulci*) were dug to a level below the foundation, assuring proper drainage.

The total width of the roads varied from 10 to 14 m, the paved carriageway was from 3 to 12 m wide. There were often unpaved side strips of 2.5 to 3 m width, called *margines* or *crepnides*. They were much used in dry weather by pedestrians and horsemen, who found the paving too hard for shoe soles studded with nails or the horseshoes of their mounts. K.D. White (1986) assumes most carriages preferred the side strips, giving a smoother ride.

The Romans used corduroy construction if a road passed through swampy marches. This was very common in Germania and gave the Romans much trouble. On his Teutoburg forest campaign Germanicus (15 BC to AD 19) sent his legate Caecina ahead, »to explore the remote forests and to prepare dams and bridges across treacherous moorland« (Tacitus Ann. I)

During their retreat Caecina was ordered to cross as fast as possible the »long bridges« (*pontes longes*) (Ann. I)

For the Via Mansuerisca in Belgium oak piles were driven into the ground, carrying timber beams and boarding, covered with stone or gravel.

We cannot explain why the Romans made their roads excessively solid, modern roads have 25 % of their thickness, carrying much higher loads. However, some Roman roads have served for more than 2000 years! The heaviest Roman vehicles (*clabula*, *sarraca* or *angraria*) weighed up to 500 kg, they had very poor suspension and since the paving was not elastic, like today's bitumen, wear and tear required oversizing. Many Roman roads have been covered by modern highways or railroads. Some were used as quarries for building material.

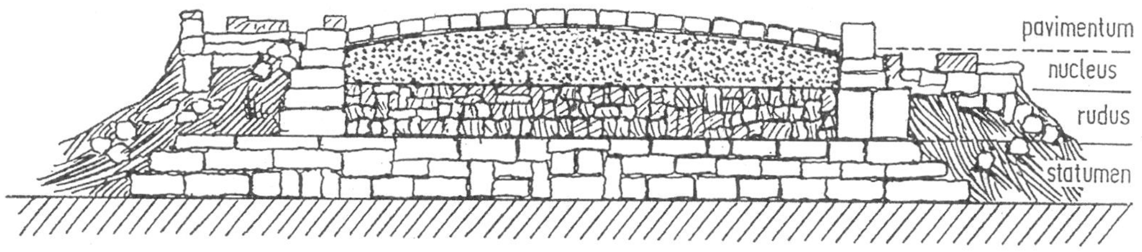

121. Typical road paving.
122. Typical Roman road section (Forbes, 1934).

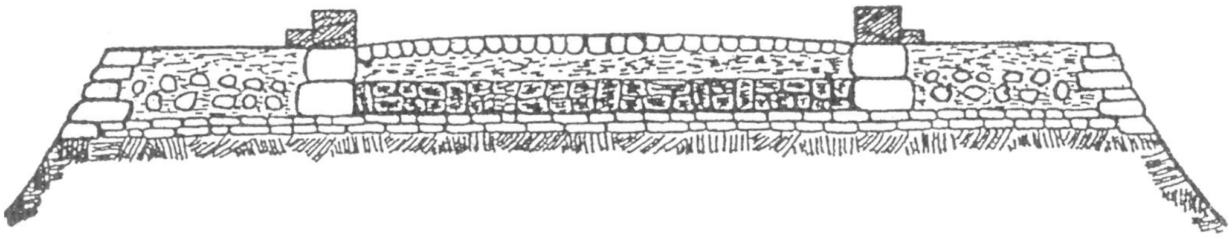

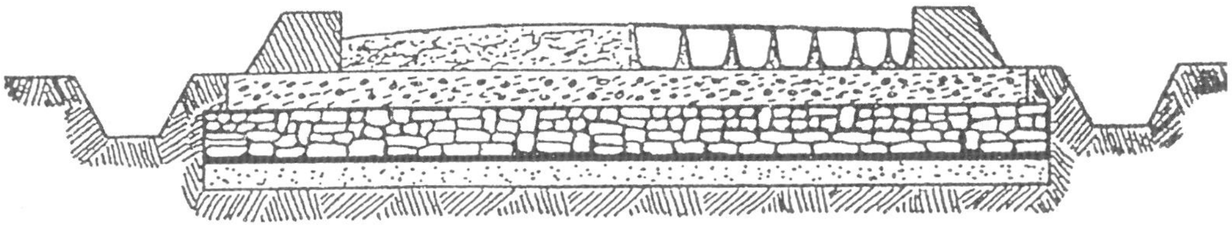

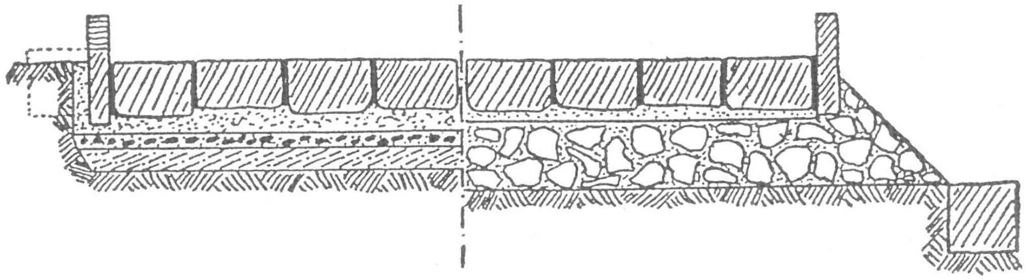

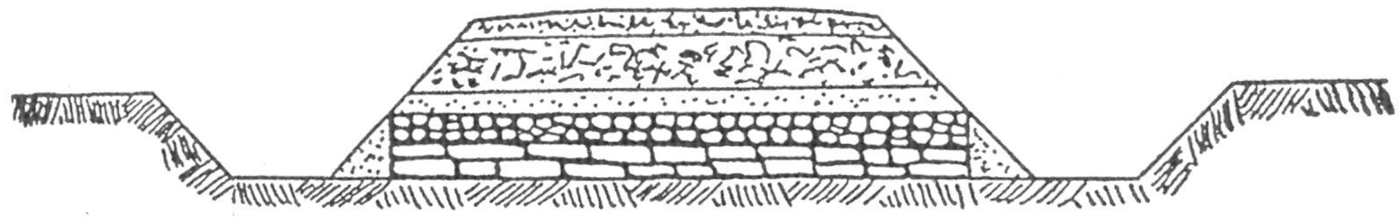

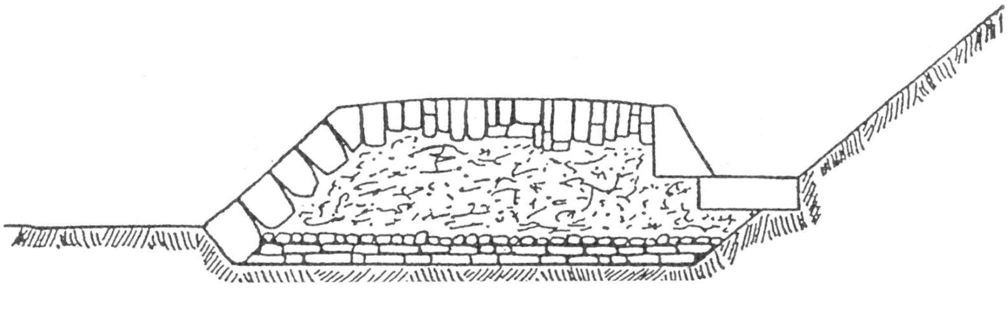

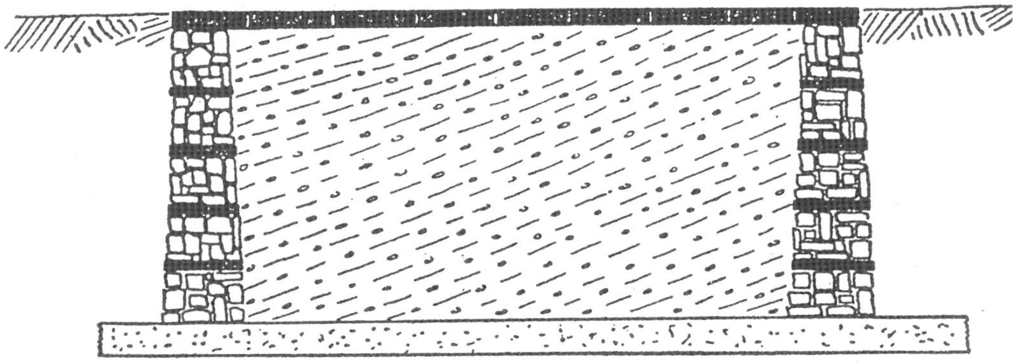

123. Highway in Italy (via publica). Pavimentum: 20–25 cm fieldstones or stone slabs laid in mortar; nucleus: 30–32 cm rubble concrete; rudus: 30–50 cm trimmed stones laid in mortar; stratum: 20–30 cm single or double layer of stone slabs.
124. Variations of above. 20–40 cm compacted gravel concrete or packed layer of stones; 30–32 cm rubble concrete and trass; 30–50 cm stones in mortar; 20–30 cm layer of sand with mortar screed.
125. Two sections of the Via Flaminia. Left: 40 x 80 cm basalt or travertine blocks in mortar, 50 cm thick; 10 x 15 cm loamy gravel; 15 cm stone chips laid in clay; 25 cm gravel concrete; c. every 5 m curbstones, projecting 45 cm. Right: same as above; 10 cm gravel in clay or loam; 60 cm broken tuff bonded in clayey gravel; cubic abutment blocks 30 x 30 x 30 cm.
126. Praetorian highway in a plain. 10 cm gravel concrete; 35 cm gravel laid in mortar; 15 cm compacted loam; 45 cm fieldstones of stone slabs in mortar.
127. Praetorian highway in hilly terrain. Heavy retaining stones and vertical paving over concrete core; base of stone slabs; stone-clad drainage channel.
128. Roman road near London, apparently built in poor ground conditions. 13 cm white clay slabs; c. 2.2 m concrete fill; retaining walls of limestone with layers of brick in between; 30 cm compacted clay; c. 2.5 m excavation.

Case study: Via Egnatia

(Based on an investigation by the Civil Engineering Department of the Democritus University of Thrace, under G. S. Xeidakis and E. G. Varagouli)

The Via Egnatia was the first Roman highway built outside Italy in the 2nd c. BC. It began in Dyrrachium (modern Durres), passed by Thessaloniki and terminated at Cypsela, east of the Evros River. Its total length was about 750 km.

From the study of sections of the road surviving in the Thrace region, it appears that the design and construction of the road was based on well-known specifications. The road was always adapted to local topography, geomorphology and ground conditions. Thus the road avoided the difficult and unstable ground, tight curves and steep grades.

The thickness and the layering of the pavement varied according to the base conditions. On stable, rocky ground, the pavement consisted of only one layer of well-fitted cobble stones. On soft and unstable ground the soft soil was excavated and replaced by several layers of cobbles, gravel and rubble, held together by compacted sandy soil or lime mortar. Up to four stone layers have been found in an archaeological excavation of the road pavement in the Thrace area. Some layers were made waterproof by well-compacted clay soil. The thickness of the pavement varied from 25 cm to more than 150 cm. The materials used were mainly of local origin.

The width of the road ranged from 4 to 8 m, depending on the ground conditions and the traffic demand. In cities its width reached up to 20 m to accommodate the increased traffic. The horizontal radius of curves was usually more than 100 m, only in mountainous areas curves of R = 10–20 m were found. The grade of the pavement normally was 1 to 2 percent, but in mountainous regions gradients of 16 to 18 % were observed. Gradients of up to 20 % were measured in a stretch of the road 2 km west of Kavala.

The cross section of the pavement was convex, with grades perpendicular to its axis from 5 to 10 %, for rapid drainage. Large rock blocks were placed at the sides of the pavement, raised above the surface, to prevent lateral spreading of the pavement and deterring carts and wagons from sliding off the road. A series of elongated stone blocks were sometimes placed in the middle of the pavement, possibly separating the opposite traffic.

5. Bridge construction

5.1. Symbolic and cultural significance of bridge building

»In my view, of all things mankind has erected and built in its drive to live, nothing is better or more valuable, than bridges. They are more important than houses, holier, because more universal, than churches. Belonging to all and being useful to everybody, always skillfully placed where most of the human needs converge – they are more permanent than other buildings and do not serve any hidden or evil purpose.

Large stone bridges, witnesses of vanished times, when man lived, thought and built differently; wind and rain has grayed and yellowed them, blunted their sharp edges, and in their joints or hardly noticeable fissures grows grass or nests a bird.« (Andric, 1971)

The bridge spanning a river is a public place, a privileged spot. Here the flow of people, of travelers, traders and soldiers, of refugees and conquerors gathers. The location changes into a focus of energy lines. Bridges connect countries and peoples. Who wants to destroy infrastructure and with it a country, must destroy its bridges.

»Temples and bridges are among the most impressive manifestations of human kind. Temples attest the belief of man in the power of God; bridges express human enterprise.« (P. Bonatz, F. Leonhardt, Cologne, 1956)

Martin Heidegger (1985) recognized the symbolism of bridges, when he writes, that we use bridges without thinking. A glance into the abyss beneath a bridge can frighten us, inducing a feeling of the risk of our existence, showing us the void over which we are balancing. The bridge spans the chasm, its ends rest securely on the ground. It translates the unconditional support of our earth into the gesture of bearing. The bridging projects over the depth into open sky, it not only connects two declivities, but suspends us in the open and is holding us. The bridge connects for us mortals earth with heaven.

Special saints flanking old bridge constructions celebrate the risk of bridge building, they give us courage and let us express gratitude for the gift of life, for this being carried between heaven and earth under their attendance. Bridges cross small lanes and multiple-lane expressways, small brooks and mighty rivers, they connect continents. They were never only examples of engineering art but also highly symbolic, as they should help to overcome alienation, hatred and prejudice between peoples, races and religions. They should lead to equality and peace.

O'Connor (1993) points out, that the word »bridge« does not exist in the Bible. Jesus says: »I am the way«, in the New Testament »the way« encompasses the whole Christian belief. Since it relates to heaven, »bridge« would have been a more pertinent term. This is very remarkable, since in antiquity rivers were associated with divine beings and the construction of bridges was an encroachment into divine territory. Cicero reports, that before fording a river, auspices were consulted (de divinatione 2.77) and the dedication of a bridge was accompanied by sacrificial rites.

During imperial times the Romans felt, they could modify nature to suit their own wishes, as stated by the Greek Rhetor Aelius Aristides in the 2nd c. AD: »You have surveyed the whole world, crossed rivers with bridges of many kinds, pierced mountains to create carriage ways, installed postal stations in empty regions, to further everywhere an orderly and civilized way of life.« As the Romans extended their rule over the whole Mediterranean region, they developed a sense of superiority over the rest of the world, well expressed by the hydraulic engineer Frontinus (de aquaeductibus 1,16): »Can you really compare the patently useless Pyramids, or those celebrated, but ineffectual works of art by the Greeks, with all those aqueducts, which serve so many people?« Pliny the Elder voices similar sentiments (nat. hist., 36,75).

Roman pragmatism expressed itself in the overcoming of natural obstacles by virtue of clear priorities, engineering skill and material prowess. They built for eternity, their structures seem to ignore the laws of decay, and the builders gained immortal glory. C. Julius Lacer, the builder of the great bridge in Alcántara (Spain), expresses this in his epitaph, calling his work *»pontem perpetui mansurum in saecula mundi.«*

The poet Statius says in a eulogy to Emperor Domitian: »Here yielded nature, conquered by the builder, to habituate itself in docile manner to an unknown purpose« (Statius, Silvae II 2)

Thus technical progress pushed religious and mysterious elements into the background, human reasoning became the basis for understanding nature. It brought us our belief in continual progress, overconfidence in engineering feasibility, overexploitation of resources and disregard for environmental change.

At all times bridges have exerted a great deal of fascination, they are among the most demanding tasks of construction, requiring courage, determination, skill and experience. They have been a favorite subject for painters, such as van Gogh, Monet, Cézanne, Munch or Turner. The esteem for bridges in our times is demonstrated by the backside design of the Euro bills. To avoid European jealousies, they do not depict real examples and as symbolic prototypes they exhibit some technical errors, which the Romans would never have committed.

5.2. Pre-Roman bridge building

Egypt

We may ask ourselves what precedents of bridge building the Romans encountered while they built up their empire. First comes to mind Egypt whose pyramids and temples the Romans greatly admired. This would go so far that they would continue fashioning temples in Egypt the traditional way. In bridge construction however, Egypt had little to offer. There hardly were any roads, as the desert could only be crossed by caravans. The Nile with its size and great fluctuation of flow could not be bridged. Indeed, it was not an obstacle but the primary means of transport. It is remarkable that no old Egyptian term for bridge has been found (Briegleb 1971).

The countless canals could be crossed with rather small bridges, most likely built with wood, of which nothing is preserved. In conjunction with the massive religious buildings very heavy loads had to be transported over short distances. There were stone bridges near the valley temple of the Chephren pyramid in Gizeh, dating from the 26th c. BC. In Gizeh exists a bridge with a greatly overdimensioned block of stone, spanning 3 m only. In Tell-el-Amara there was a 5 m wide timber bridge on stone piers, spanning 3.5 and 5 m. Briegleb mentions a couple of pictorial records of bridges in the temple of Sesthos I in Karnak, 12th c. BC. A relief depicting the campaigns by Ramses against the Hittites shows a nondescript bridge and we know of a 14th c. BC Hittite bridge at Boghazköy, the capital of their kingdom, only 1.7 m wide, but spanning 8.5 m about 14 m above the river. Two cantilevering stone slabs carried another slab with 6.5 m clear span.

Mesopotamia

Mesopotamia was florishing between the Tigris and Euphrates Rivers thanks to countless irrigation canals, which had to be crossed. Being one of the oldest civilizations, we may assume that bridges were built at a very early time. The first mention of a bridge is in an inscription of King Addapaksu of Susa, 3rd millennium BC. Written records indicate wood- and pontoon bridges from the 2nd millennium to the 9th c. BC, but stone bridges are not mentioned until the time of Sargon II (722–705 BC) and his son Sanherib (705 to 681 BC).

Briegleb writes about a bridge at the citadel of Khorsabad built by Sargon II: 15.9 m wide and 5.1 m high it had a pointed arch of 3.7 m span. Sanherib built an bridge at Djerwan, 22 m wide and 28 m long. With a height of 4.9 m it consisted of five pointed arches of 2.8 m span each. A bridge at Nineveh of the same time is shown on a relief. Construction details are not known, but the depiction of an aqueduct at Nineveh suggests false pointed arches by corbeling. From inscriptions of the 6th c. BC we know of a bridge across the Libilhegalla Canal at Babylon.

One of the most famous bridges of antiquity was the bridge over the Euphrates at Babylon. It is described by Herodot (ca. 485–425 BC), Curtius Rufus and Diodor (both 1st c. AD). Connecting the two parts of Babylon it is the oldest known stone bridge (Wölfel 1990). In 1910/11 the German archeologist R. Koldewey studied the remains of the bridge. He located two abutment piers and six piers in the river (R. Koldewey, Leipzig 1925). Total length was 126.4 m, against eight stadia given by Diodor (= 1500 m). The width of the piers varied from 23.8 to 11.8 m at the base, tapering to 21 to 9 m at the top. They were variably spaced at 7, 8.5, 4 x 9 and 19.4 m and had a boat shaped cross section. Most likely the bridge itself was made of timber, the big span possibly being a kind of draw bridge, to afford passage of tall ships.

Herodot gives a lively description: »The town constited of two halves, separated by the river. Who in previous times wanted to pass from one part of the town to the other had to take a boat, which was not convenient. Nitokris again brought help. When that lake was dug, they used it also for the following memorable construction. They prepared big stones and when everything was ready for construction, they diverted all the water of the river into that excavated basin and while the basin filled up the river went dry. Then they stabilized the river banks and the steps leading to the water along the town with brick masonry, similar to the city walls. Further they built in the middle of the town a bridge in stone masonry with the excavated stones and connected the stones with iron and lead. During the day they placed wooden beams across, so that the people of Babylon could cross the river. At night they removed the beams, to prevent people from crossing and stealing from each other. After completion the Euphrates was redirected from the filled lake to its old bed."

It is striking that all three authors write of »dressed stones« connected with iron and lead. The excavations clearly showed piers built with 31 x 31 cm bricks laid in asphalt mortar.This required neither metal inserts nor a stone facing. Diodor (II 8) reports that the piers had cut waters and that the timber parts were of cedar or cypress wood. He bases his report on a description by Ktesias of Knidos (about 400 BC), who spent a long time in Persian custody and must have known the bridge from personal observation. His main works deal with India (Indika) and Asia Minor plus Persia (Persika). Curtius Rufus stresses the difficult foundation works. Indeed, the diversion of the river was the only method to carry the foundations down to bedrock.

The dating of the bridge can be done as follows: Near the eastern abutment pier was an embankment wall built by King Nabupolassar (until 605 BC). Against it the first pier was built by King Nabonid (555–536 BC). Accordingly, the bridge must have been built after the death of Nabupolassar and before King Nabonid, which means during the reign of Nebuchadnezzar II (605–562 BC).

Crete, Mycenae and Greece

With the exception of northern Greece there are hardly any permanent rivers requiring stone bridges. The many autonomous city states did not lead to a network of roads. Since most settlements had access to the Mediterranean, trading was done via the sea. This is in contrast to the centralized Roman Empire, which ruled via its extended road system. Lacking sufficients remains it is impossible to give a satisfactory picture of bridge building in this region.

Most ancient is a bridge at the palace of Knossos, 2nd millennium BC. The bridge is a part of a long approach. Width 5.25 m, height 4 m, three spans from 2.3 to 3.25 m. Built by corbeling the openings had trapezoidal or triangular shape. Briegleb mentiones a bridge over the river

Chavos south of the Citadel of Mycenae (1400–1200 BC). It was part of a 50 m long, 5.84 m wide and 4.5 m high dam. A cyclopean bridge of the 14th c. BC can be found near Kasarma on the road from Nauplion to Epidauris. At the entrance to the village of Arkadikó there is a 1.4 m opening in a 5 m wide and 3.8 m hight dam. It is still in use.

In Greece proper there is mention of a fourth century BC bridge on the sacred way from Athens to Eleusis, built by a certain Xenokles. An excellent example of antique engineering skill is the bridge at Assos in western Turkey. A river bed of 50 m width was flooded sporadically. The permanent flow was crossed by a 13 m timber construction, while the rest was managed by a succession of 17 stone slabs, 34 cm thick and spanning 3 m on 2.5 m above flood level supports. It was built during the 4th c. BC. The earliest known use of an arch bridge from hellenistic times is found on Samothrace at the Kabir's sanctuary. Here stood the famous Nike of Samothrace. Along the processional street a brook had to be crossed. A vault of five voussoirs spans a mere 1,9 m with a depth of 13 m. Most likely it was built after 279 BC by Ptolemaios II (283–247 BC).

An absolutely special case presents the triple bridge at Meligala in Messenia. Three bridges meet at 120° angles, to cross the rivers Maurozoumenos (*Leukasia*), Vasiliko (*Amphitos*) and a low aproach wing. The bridge has been maintained and rebuilt until our time. Few parts consist of the original squared masonry from antiquity, mainly up to eight ranges of the central pier. The bridging consists of typically Turkish pointed arches with spans from 5 to 7 m from the 16th c. AD. It is thought that the original bridging was done by corbeling.

There are several other bridges built by corbeling:

Bridge in the Laurion mountains near Athens.

Bridge at Xerokambios in Laconia.

Bridge at Agrigent from the 5th to 4th c. BC, 1 m span.

Two similar bridges at Eleutherna from the 4th c. BC, one of them still in use. The 3.8 m spans are tackled by corbeling, the top part forming an isosceles triangle of 1.83 m height. Length 9m, width 7–8 m, 6 m height 6 m.

Bridge at the westgate of Eretria, 500 BC. Wood or stone slabs on stone piers. Length 6.27 m, width 5.4 m, height 2.8 m, span 3 m.

Stone slab bridge at the Artemis sanctuary od Bauron, 5th c. BC. Length 9.2 m, width 9 m, span 1.4 m.

Bridge in the battlements of Syracuse, length 8.2 m, width 6.8 m, height 8.5 m, span of two openings 2 m each.

Persia

The most famous Persian bridge operations were the floating bridges of Darius I (522–486 BC) and Xerxes I (486–465 BC) across the Bosporus and the Dardanelles during their campaigns against Greece. Of course, those sannot qualify as bridges in our sense. The highly centralized Persian Empire also relied on a road system requiring bridges. However, there is very little information, again showing that bridges did not draw any attention. Darius I built many water management structures, which also featured dams with sluice gates. The gates were bridged over since the dams were also used for traffic. Such structures existed near Persepolis across the Kor River, the piers were of large stone masonry (Harting 1972).

The sluice gate bridge of the Karun dam at Shushar already shows Roman technology. It was built by the Sassanids under Shapur I (AD 241–272) after his victory over Emperor Valerian in the battle of Epudar (AD 260). He captured about 80,000 Roman prisoners, among which there surely were engineers and artisans. They were employed for road, bridge and irrigation works. The 500 m bridge with 40 arches is a purely Roman structure with semicircular vaulting. The 15 m wide dam rested on bedrock and and dammed the river up by up to 4 m.

5.3. Roman stone bridges

The chapter on bridges shall give a comprehensive review of Roman stone bridges, taking into account the historical background, archaeological studies, literary sources and technical conditions. Our study is necessarily limited to bridges or their piers, built in stone or masonry, since nothing remains of wood as a material. The construction of timber or pontoon bridges is known from accounts by antique authors, or through some remains of pile foundations. Well known examples are the Mosel bridge at Trier and the bridge over the Danube at Turnu Severin in Romania, built by Trajan during his campaign against the Dacians. His engineer Apollodorus of Damascus put the whole army to work for a whole year.

We have today at least 400 Roman stone bridges in complete or partial existence. It is difficult to give an exact figure, many bridges contain only small portions of Roman origin, having been reconstructed through the course of centuries, after destruction by warfare or natural catastrophes. Of some bridges we can only assume that they are built on Roman foundations – they hardly qualify as Roman.

Through the Middle Ages till modern times the semi-circular arch with radial voussoirs has been employed for bridge building. Many later bridges therefore appear on first sight as Roman. Only detailed study can tell us if they are originally Roman. Their state of conservation is not a clear criterion, since Roman bridges often are better preserved than those from the Middle Ages. Many Roman bridges continue being used to this day, since they are continually maintained and repaired and are able to take today's heavy traffic.

In medieval times existing Roman bridges served as model for some large stone bridges, such as in Regensburg, Prague or Avignon.

During the Renaissance, which rediscovered antiquity, Roman bridges also became items of interest. Palladio (1508–1580) measured up a Roman bridge in Vicenza and presented it in his *quattro libri dell'architettura*.

If we take into account that there were about 150.000 km of principal highways in the Roman Empire, with another 150.000 km of secondary roads, we can assume the existence of thousands of bridges. Since the best way to study Roman bridges is on site, the author has visited and measured up more than 250 bridges, still completely or partially standing. In many cases he could ascertain their extraordinarily good condition, the result of superior workmanship.

The names of existing bridges date mostly from later times. They refer to their location or derive their name from legends about the circumstances of their construction. Very frequently they are called »devil's bridge«, in particular if they are of daring construction, suggesting the devil's involvement.

More and more Roman bridges are becoming redundant through new construction, they deteriorate because of failing heritage agencies, their misappropriation as stables, stores or dumps, or just by mindless neglect. Further loss of these unique structures must be prevented. While other monuments of antiquity are generally well cared for, bridges are mostly overlooked. This study hopes to further the recognition of Roman bridges as unique technical monuments.

5.4. History of Roman bridge building

Roman bridges can be categorized by different phases of technology and execution, each with its individual style (W. Müller, 1989). They correspond to the political periods as follows:

146 to 31 BC	Late Republic
31 BC to 14 AD	Time of Augustus
41 to 68 AD	Time of Claudius and Nero
69 to 96	Flavian period
96 to 117	Time of Trajan
117 to 138	Time of Hadrian
138 to 192	Period of the Antonines
192 to 235	Severian period
235 to 285	Time of Barracks emperors
284 to 363	Diocletian and Constantine period
364 to 455	Valentinian and Theodosian period
455 to 476	Last West-Roman emperors

5.4.1. Known bridges and their remains

Piero Gazzola has listed in his work *Ponti Romani* (Florence, 1963) 293 Roman bridges with brief descriptions. This compilation is not at all complete and partly superseded. Many bridges – mainly outside Italy – are not included or do not exist any more. Since the Roman network of roads was most dense in Italy, we find here most of the Roman bridges still standing. According to Gazzola the Roman bridges listed by him are distributed as follows:

60 %	Italy
10 %	Spain and Portugal
10 %	France, Belgium, Germany and Austria
10 %	North Africa
10 %	Turkey, the Balkans and the Near East

Today's condition of the bridges depends upon the quality of execution, the material used, the location and local climate, the ease of maintenance – or its utilization as a quarry, the impact of natural catastrophes such as earthquakes and floods, and of wars. Particularly in Italy a large part of Roman bridges was destroyed during World War II.

129. Gatehouse in the middle of Ponte Cecco at Ascoli Piceno, Italy.
130. Triumphal arch in the middle of the Alcántara Bridge in Spain.
131. Aedicula for commemorative tablet on Augustus' Bridge in Rimini, Italy.
132. Two triumphal arches adorning the Pons Flavius at Saint Chamas (Bouches du Rhône, France).
133. Gate over the Via Domitia at Glanum in the French Provence.

134. Small temple at the eastern embankment of the Tejo bridge at Alcántara (Spain). In the inscription on the marble slabs over the entrance, Gaius Julius Lacer declares to have this bridge built for eternity (*pontem perpetui mansurum in saecula*).
135. An inscription on the triumphal arch of the same bridge lists the cities contributing to the construction costs: *municipia provinciae Lusitaniae stipe conlata quae opus pontis perfecerunt.*
136. Exedra at the east end of Justinian's Bridge in Adapanzari (Turkey), originally adorned with statuary.
137. Inscription over an arch of the Pons Flavius at Saint Chamas (France) telling us, that Lucius Donnius asked his inheritors, among them his son and his brother, to commission the construction of this bridge (CIL XII, 647).

5.4.2. Antique sources of information

Many bridges carry inscriptions, reliefs or sculpture. They allow deductions of the time of construction, the sponsor, the importance of the project and other details. Often the inscriptions name the architect (then the equivalent of engineer and contractor). There are even details of funding.

There is an inscription on the Pons Fabricius in Rome (62 BC): »Lucius Fabricius, son of Caius, in charge of road construction, had this bridge built and accepted it« – *L(ucius) Fabricius C(ai) f(ilius) cur(ator) viar(um) faciundum coeravit eidemque probaveit* (CIL I, 751). A second, smaller inscription refers to repairs after a flood in 23 BC: »Marcus Lollius, son of Marcus, (and) Quintus Lepidus, son of Manius, received as consuls, by decision of the Senate, again this bridge« – *M(arcus) Lollius M(arci) f(ilius), Q(uintus) Lepidus M(ani) f(ilius con(n)s(ules) ex s(enatus) c(onsulto) probaverun(t)* (CIL I, 751).

Beyond the bridges themselves we find inscriptions on monuments (e.g. Trajan's column in Rome) or as information in official accounts (e.g. the »res gestae« by Augustus on a temple in

Ankara). Arched gateways may be in the middle or at the end of a bridge, or marking a major highway with relevant inscriptions. Sometimes a bridge was dedicated to a god, including altars, an exedra or even a small temple.

Further sources are coins with information about road or bridge construction. To be mentioned a bridge over the river Maiandros (today Büyük Menderes), south of today's Nazilli near Antiochia in western Turkey. This bridge figures on many imperial coins from this city.

5.5. Design criteria for bridges

1. Smooth continuity of highway, maintaining levels, width and carrying capacity.
2. Maximum water level, quantity and speed of flow.
3. Condition of terrain (stability, bearing capacity).
4. Availability of building materials (quarries, sources for timber and metal).
5. Required quality and strength of materials.
6. Availability of workers and machinery.
7. Financing.
8. Purpose and serving life of bridge. (In 55 AD Caesar made do with a light, temporary structure for a bridge over the Rhine, since he planned a limited campaign.)
9. Available construction time.
10. Available technology and know-how.

5.5.1. Service loads

Some of these points have been discussed in chapter on road construction, as they apply to all building activity. Bridge specific is the question of the service loads being produced by pedestrians, animals for riding, bearing and pulling, and the weight of two- or four wheeled carts and carriages. There is some information about the construction of such vehicles in fragments of Diocletian's decree on prices (*edictum de pretiis,* AD 301).

K. D. White (1986) presents a few models. Two-wheeled carts such as the light and fast *cisium* or the more comfortable *birota*. Among four-wheeled carriages was the open *reda* and the covered *carruca*, which could be transformed into a *carruca dormitoria*, a sleeper. The term *carpentum* was also used, but applied for government transport, mainky for the imperial family. For transporting goods served the light *plaustrum* or the heavy *sarraca*.

The wheels of light vehicles had eight spikes and rims of six segments. Suspension was primitive: rigid axels attached to bearings below the vehicle. Elastic suspension by means of leather straps was not known. Carriages for heavy goods had solid disk wheels, allowing loads of up to one ton. Oxen to pull them would attain a speed of 3 km per hour, while travelling carts covered more than 50 km per day.

138.Loading of the Roman bridge over the Rhine at Koblenz according to H. Fehr (1981). Farbahnlast = carriageway load, Tragwerkslast = weight of structure,
139. Relief of a two-wheeled *cisium*, Trier, AD 200.

140. Relief from Trajan's column in Rome. The emperor makes an offering in front of the bridge over the Danube. Above we see piers and the wooden truss bridging.
141. Partial view of the bridge (Wölfel, 1990).
142. Artist's view of Caesar's 400 m long bridge over the Rhine, 55 BC (Köthe, 1991).
143. Reconstructed elevation of the Danube bridge (Müller, 1989).
144. In AD 310 Constantine the Great built a 10 m wide and 420 m long bridge over the Rhine at Cologne. It had 15 piers at variable distances, the span of the wooden trusses averaged 26 m. In the 9th c. the bridge was dismantled. Reconstruction of typical bays by Menschig, 1981.
145. Fancy reconstruction of Trajan's Bridge over the Danube at Turnu Severin, disregarding the actual number of piers (Cichorius, 1900).

The best proof of the superior strength of Roman bridges is the fact that so many have not only survived two millennia, but were able to support whole columns of panzers during World War II, such as the Pons Milvio in Rome.

5.5.2. Timber bridges on stone piers or wood piling

Whenever speed was important, or it was for temporary use during a campaign, the Romans built with timber. Naturally, nothing remains of such bridges, save the foundations. We have to rely on descriptions and pictures for information. Caesar reports on the bridge over the Rhine, built around 55 BC in the space of 10 days. Transverse girders on twin piles carried longitudinal 10 m beams, which were covered with wattle and soil. (Gallic War, 4, 17–19.)

Longer spans required truss frames, often in polygonal form. The best known example is Trajan's 1135 m long bridge over the Danube at Turnu Severin. It had 26 stone piers and spans up to 50 m. It was completed by the military within a year. Trajan's column in Rome depicts the structure. Hadrian had the timber part removed, to stop the Dacians from crossing the river.

In 310 AD emperor Constantine built a 420 m long bridge over the Rhine at Cologne. It rested on pile grills, had a width of 10 m and spans averaging 26 m.

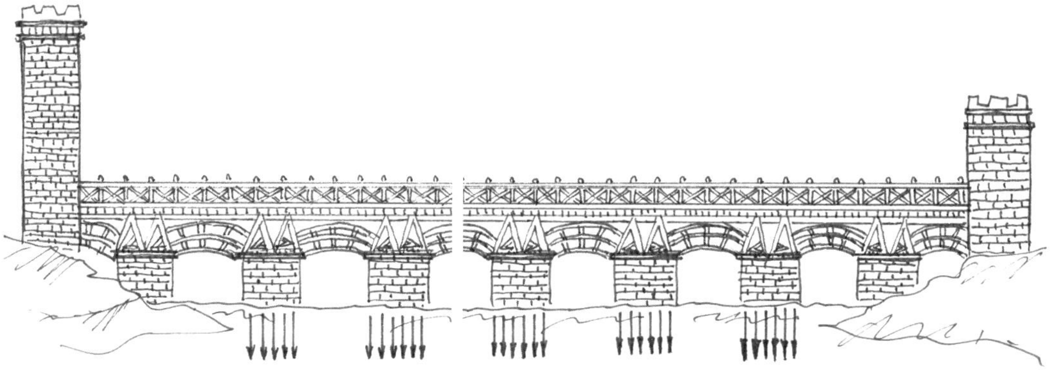

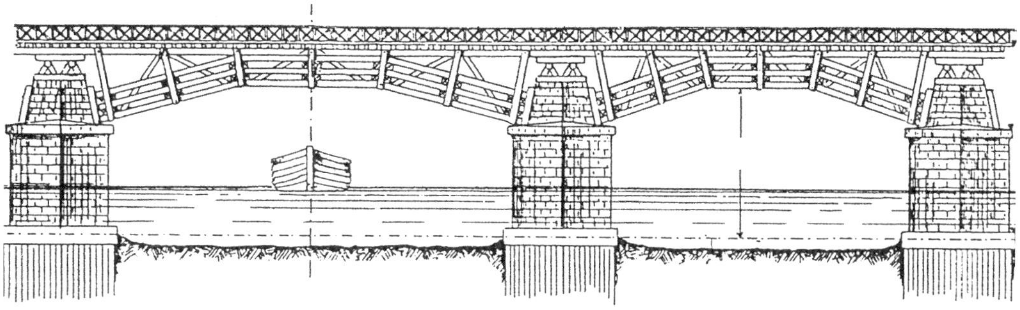

5.6. Structural systems for stone bridges

Our study excludes aqueducts. *Pontes* meant for the Romans exclusively the crossing of a river for human traffic (*iter super fluvium*). The famous Pont du Gard combines aqueduct and road bridge.

146. Stone slab as simply supported beam.
147. Sketch of stone beam at Postbridge.
148. Bending diagram of a beam on two supports.
149. Elevation of a »clapper bridge«.

5.6.1. The simply supported beam

The oldest and simplest form of bridging is a beam supported on either end. The possible spans depend on the strength of timber or stone. Subjected to a load the beam will bend, the upper part being subjected to compression, the lower part to tension. The middle is neither under tension nor compression and is called the neutral zone.

Since stone is not elastic and weak in tension, a stone beam with large span may fail under its own weight, developing a fissure on the underside. Spans of bridges made of stone slabs are therefore limited to 4 m. But stone is very strong in compression, granite up to 210 N/mm^2, dense limestone up to 180 N/mm^2, while good concrete reaches today 80 N/mm^2. Such values are never reached in construction – a granite column could reach a height of 6000 m before crushing under its own weight, limestone may reach half that height.

Timber beams can have a far longer span, because compressive and tensile strength are more balanced. However, the span is usually limited to 10 m, because beyond this length deflection would be the determining limit and the beam would start oscillating under a moving load.

A typical example of a flat stone bridge with two supports is the Clam bridge in Wycoller (Lancashire, England) dating from 1000 BC. It consists of a monolithic stone slab, a *clam* on two rock embankments.

A similar bridge of unknown date exists at Postbridge (Devonshire, England), where a heavy stone slab rests on two granite bases at a distance of 2.5 m.

Further developments are the so-called clapper bridges, consisting of several spans in a row. There is again at Postbridge a good example, with two granite slabs, each 4.5 m long by 2 m wide and 30 cm thick.

In Somerset the so-called tarrsteps were built around 1000 BC, a sort of causeway having many small openings bridged by 24 stone slabs. It was about 40 m long and 1.5 m wide, the slabs weighed up to 5 tons.

The biggest slabs of such bridges reached a weight of 12 tons, entailing a serious transport and erection effort. Stonehenge shows, that the builders of the Bronze Age were capable of such feats, using 30 ton blocks from 30 km away.

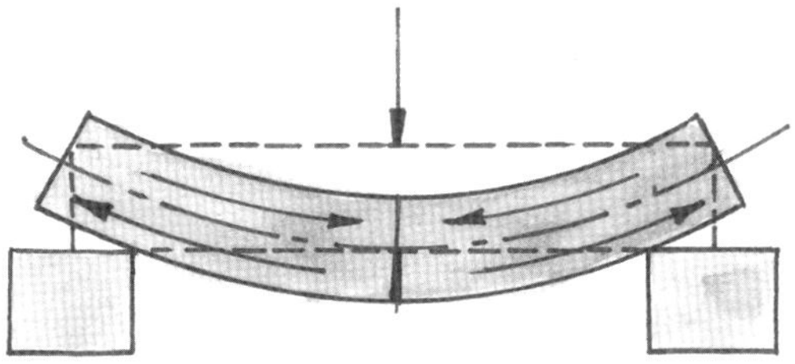

148

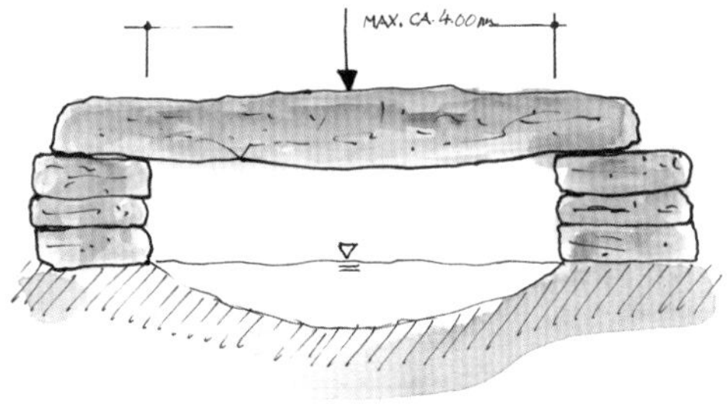

146

147

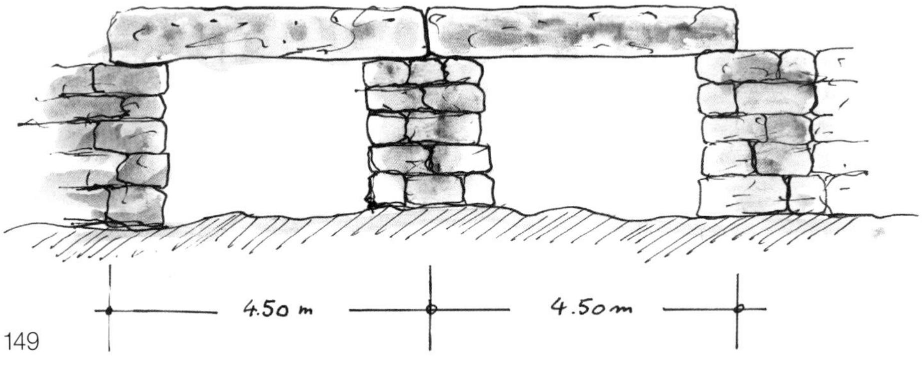

149

150. Passage through a wall at Kaunos, Caria, Turkey.
151. Bridge at Selinent, current condition of bridge with wooden deck.
152, 153. Stages of corbelling.
154. Corbel bridge at Selinunt (Sicily) as part of a fortification. It leaves a gap between the top stones. Possibly, the gap was bridged by timber beams, which could be quickly removed in case of attack.
155. Completed corbelled (false) arch.
156. Bridge at Eleutherna in Crete (4th c. BC). It is 9.5 m long, 5.4 m wide and 4.6 m high and has a triangular opening 3.8 m wide. If the top would be formed by a flat slab we would have a trapezoidal shape, or the corbels could be trimmed in a curved fashion, giving the impression of a pointed arch (Heinrich, 1983).
157, 158. Bridge near Cesme Köy, western Turkey, 4th c. AD. It crosses a 70 m wide valley on a causeway, built of two ashlar quarry faced stone walls, 7 m apart, with a rubble core. It had a 3 m gap for the stream. Only the down stream part remains, most likely the rest was washed away. Curiously, the eastern part has regular 29 cm high ranges, while the western wall has variable ranges, up to 59 cm high. Access from the east is sunken to reduce the height of the bridge to about 4 m. The excessive width of the bridge may be explained by the desire to emphasize the importance of connecting Old and New Knidos.

5.6. 2. Corbelling method

The corbelling method allowed an increase of span, using relatively short stone slabs. Each layer is slightly projecting, keeping its center of gravity well within the stone below. The weight of the wall behind keeps the stones in place and once the top is closed we reach the state of a false arch. To enhance this effect the stones could be trimmed back to give the opening a triangular shape, also reducing the weight and allowing larger projections.

Corbelling is an ancient method of producing arch-like or vaulted structures without temporary shuttering. It is typical for mud brick construction and was widely used in Mesopotamia and Egypt.

152

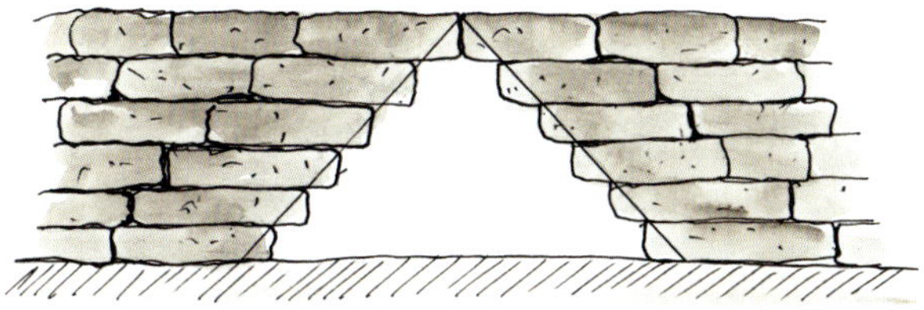
155

153

156

154

150

151

157

158

5.6.3. Evolution of arches

The inclined positioning of beams produces a false arch. If we lean two beams against each other without fixing them at the bottom and put a load upon them, the beams will be pushed apart. Following the diagram of forces the load will be split up into horizontal and vertical components. The horizontal push can be absorbed by buttressing or a tie rod (as introduced by Italian Renaissance builders).

It seems, true arches have been known by the Assyrians and Babylonians since 3500 BC. In Ur a king's tomb dating from the 4th c. BC has been excavated, featuring a gate and vaulting with semicircular brick arches. In Egypt true arches have been employed since the 3rd millennium BC, and the Greeks did so in classical times, but in all these cases such construction was limited to subterranean work, to ensure sufficient buttressing.

An Etruscan bridge with stone voussoirs near Viterbo indicates, that they were the Roman's teachers, as in so many other things (Lugli, 1957). Examples are the Ponte della Rocca near Biedia (Etruria) or the Ponte della Badia at Vulci.

After the subjugation of the Etruscans by the Romans, their high standard of living and their skills were played down by their new masters.

Seneca (Epist., 90, 32) writes, that Democritus of Abdera (c. 460 BC) was the inventor of true vaulting. Most likely, the fame of Democritus as universal scientist prompted this view.

5.6.4. True-arch construction

The height of corbelled structures had a practical limit of 5 m, anything beyond became too bulky. The problem was to find a technique, which would fully exploit the superior compressive strength of stone. The solution was the use of wedge-shaped stones, arranged in a curved fashion. It imposed two new requirements: Sufficient timber for the temporary shuttering and solid buttressing to resist the sideways push of the arch. The everlasting horizontal thrust of an arch is well expressed in the Arab saying: »An arch never sleeps.«

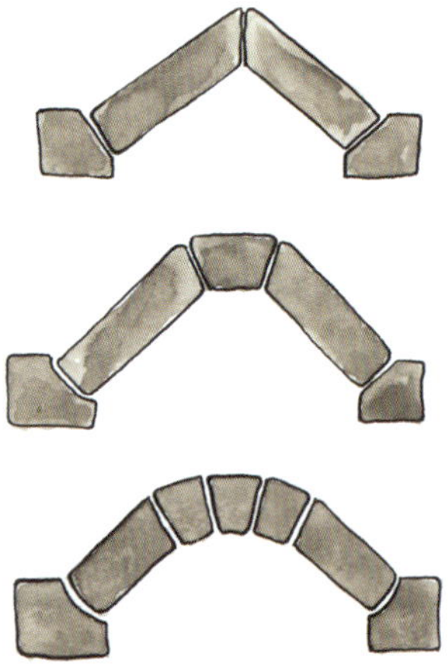

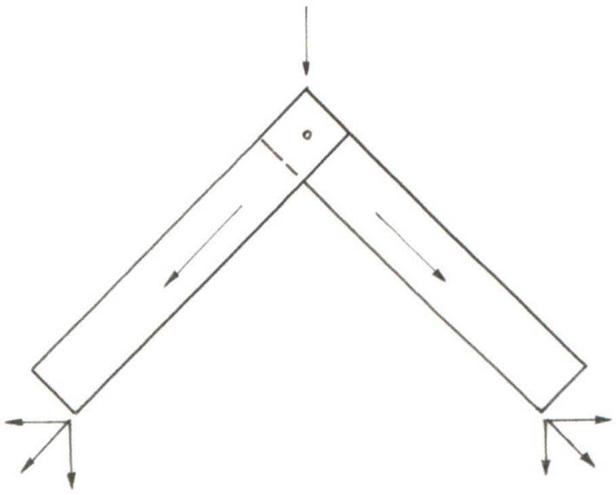

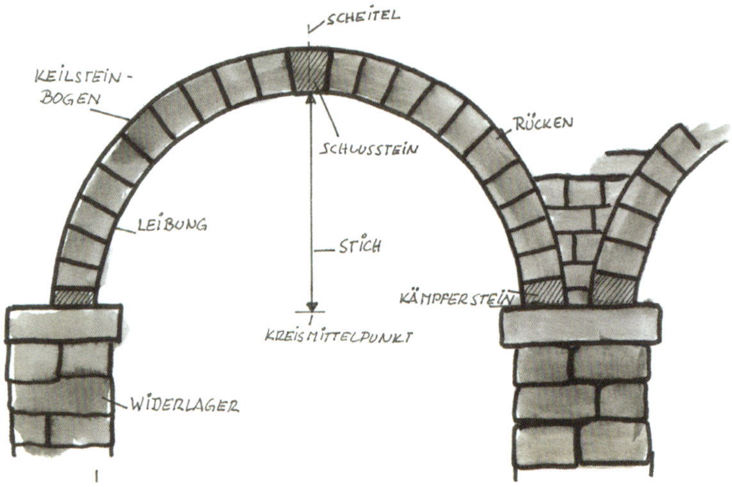

159. From the false to the true arch.
160. Two inclined beams hinged at the top divert a vertical load.
161. Arch construction using wedge shaped stones.
162. Ponte de la Badia at Vulci (Italy) of Etruscan time.

163. The wedge stone solution.
164. Nomenclature of arch components (A = abutment, S = springer, V = voussoir, In = intrados, Ex = extrados, K = keystone, I = impost, P = pier).
165. Semicircular arch of a tomb at Lydai in Caria (Turkey), about to fail due to lack of buttressing. Only friction is keeping the three central voussoirs in place.
166. Flow of forces in arches and rule-of-thumb dimensioning. (Schlussstein = keystone, Stützlinie = pressure line, Kämpfer = impost, Spannweite Sehne = span).

A true arch consists of wedge shaped stones, called voussoirs, which are arranged in a semicircle with radial joints. By their radial arrangement the voussoirs brace each other and keep them in suspension. The number of voussoirs should be uneven, to provide for the last stone at the top, which is aptly called keystone and completes the arch. Vertical loads are deflected along the curvature of the arch, starting horizontally at the top. The resulting push must be arrested by the weight of the spandrels and abutments. A single arch usually has the width of a beam and is a linear support. If we place many arches side by side we receive a tunnel-like structure, called a vault. The Roman engineers employed both methods to perfection.

Heyman (1969) devised a formula to determine the minimal height of voussoirs for a given span of a bridge under its dead load. The values differ for various segments of a circle:

Semicircle 180°	H/R = 0.1060
Circle segment 160°	H/R = 0.0680
Circle segment 140°	H/R = 0.0410

We can see that a reduction of the segmental opening diminishes the voussoir height or allows a larger span. The values presuppose a perfect circular arch shape. Deviations caused by sagging abutments, poor fit of the voussoirs, imperfect condition of the falsework, may require an increase of voussoir height by up to 25 %. It is also assumed that the superstructure of spandrel and abutment walls will add to the strength of the masonry, as would a core of concrete.

The measurements of about 100 bridges gave a mean value of H/R = 0.2150. The smallest values were found at the Eski Kahtâ Bridge in eastern Turkey = 0.0351 and St. Martin Bridge in Italy with 0.0382.

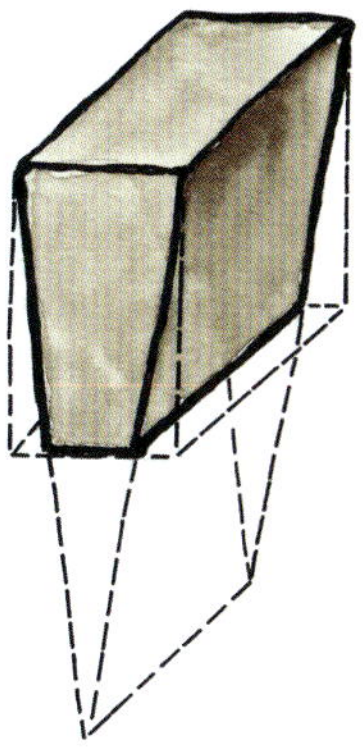

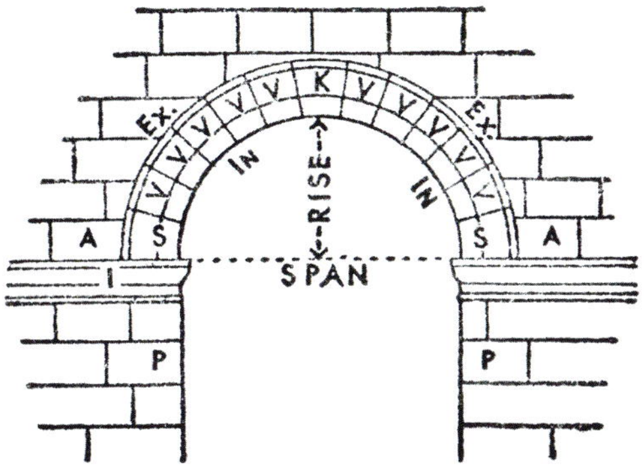

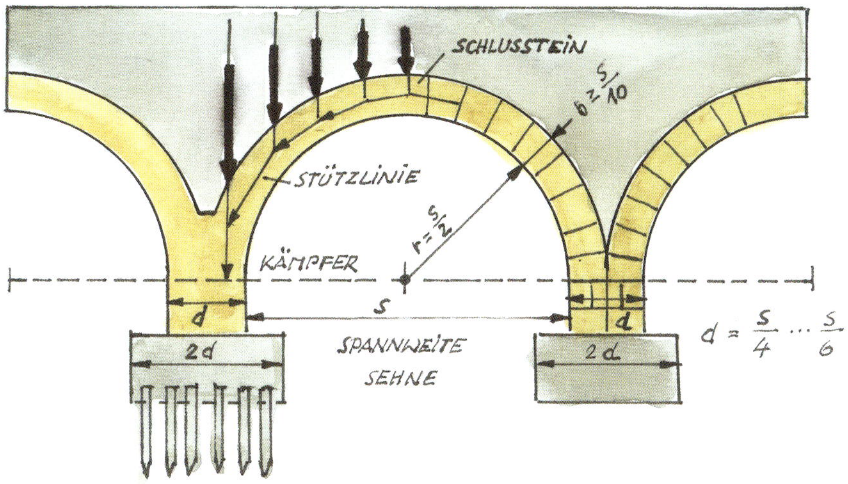

167. Examples of Roman and Etruscan arch shapes (Lugli, 1957).
168. Example of arches as substructure of a dome over a square space. The pendentive (= part of a circumscribed sphere) between the arches produces a circular base for the dome.

169–171. Combinations of stone, brick and concrete.
169. Marsias bridge at Gergas (Turkey).
170. Val Quazzola bridge at Quiliano (Italy).
171. Val Ponci bridge at Finale (Italy).
172. Bridge at Toruncova (Turkey).
173. Ponte di Augusto at Narni.
174–177. Examples of stretcher bonds for vaults (O'Connor, 1993).
178. Iron clamps in the Cestius Bridge in Rome.
179. Various materials for vault construction, from the top down: brick, dressed stone, field stones, concrete.

The listing of bridges by date of construction reveals a steady reduction of Heyman's value due to increasing experience of the builders:

Period of construction	Number of bridges	Ratio H / R
2nd c. BC	8	0.3690
1st c. BC	41	0.2010
1st c. AD	16	0.2030
2nd to 6th c. AD	29	0.1850

The ratios H / R for individual bridges are given with the Case Studies

Structurally a vault consists of parallel arches, having tie stones to link with each other for improved cohesion. Often individual vaulted sections were built in succession, repeatedly using the same shuttering by shifting it. Critical sections of masonry were reinforced with iron clamps set in lead against corrosion.

174

175

176

177

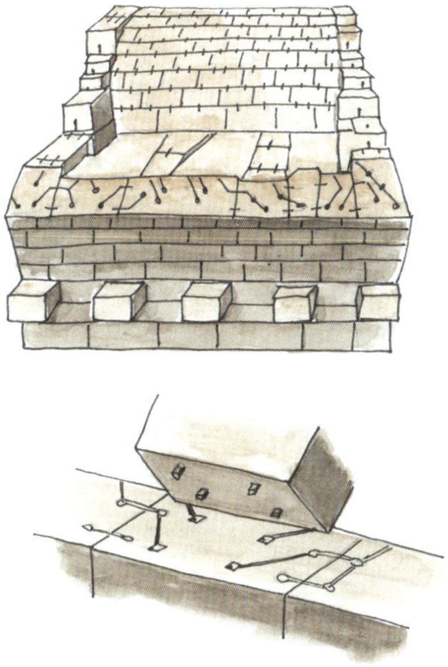

178

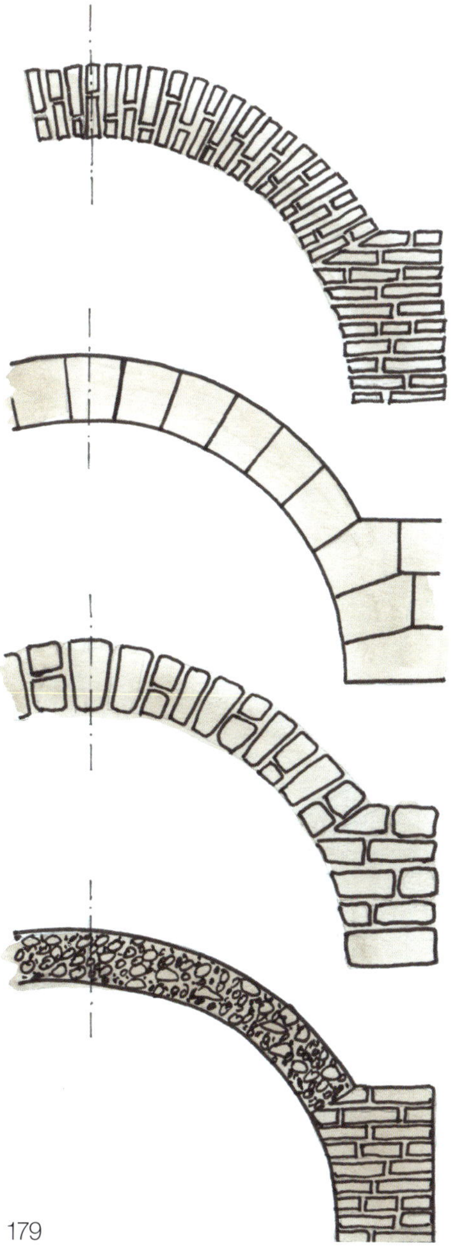

179

The Romans have been criticized for limiting themselves to semicircular arches. Since structural analysis was not known, they were not aware that parabolic shapes were much better following the pattern of forces. If we suspend a chain from the two ends it will assume a catenary line, which follows the natural flow of forces. Taking this form vertically, we arrive at a catenary arch which is in full equilibrium, generating loads of compression without any friction from one voussoir to the next. These forms are used in modern concrete construction for slender arch-bridges or thin shell construction.

In 1695 Philippe de la Hire developed a theory of vaulting along these lines and in 1717 the English mathematician James Stirling devised a vault with spherical units, which transmitted pure point loads according to the »chain line«.

The disadvantage of deviating from the circle was the complicated execution, every voussoir having a different inclination. Segmental arches used during late imperial times are a good compromise since the upper part of a true arch comes close to catenary condition. Examples are Justinian's Bridge at Adapazari (Turkey) and the bridge at Alconétar (Spain).

180. A genius like Michelangelo employed a parabolic shape for St.Peter's cupola (Poleni, 1748).
181. Catenary chain and catenary arch, being of near parabolic shape (Büttner/Hampe, 1977).
182. Segmental arch.
183. Top Part of parobola.

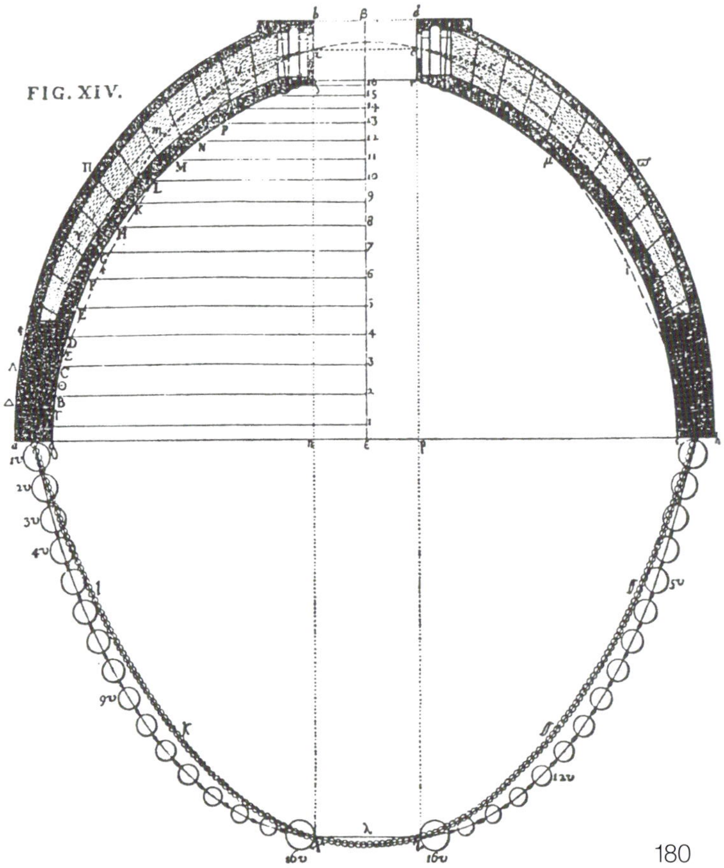

180

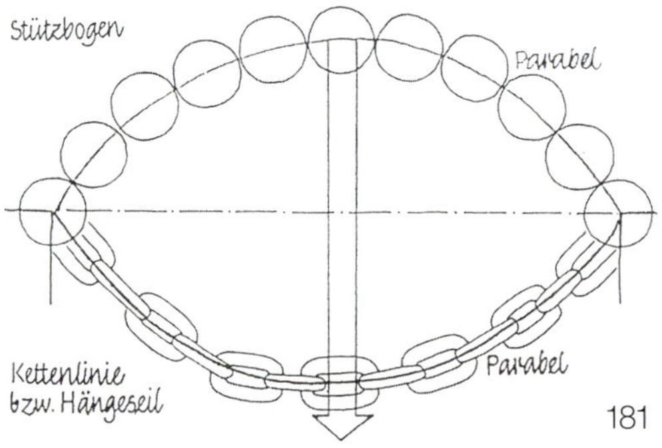

181

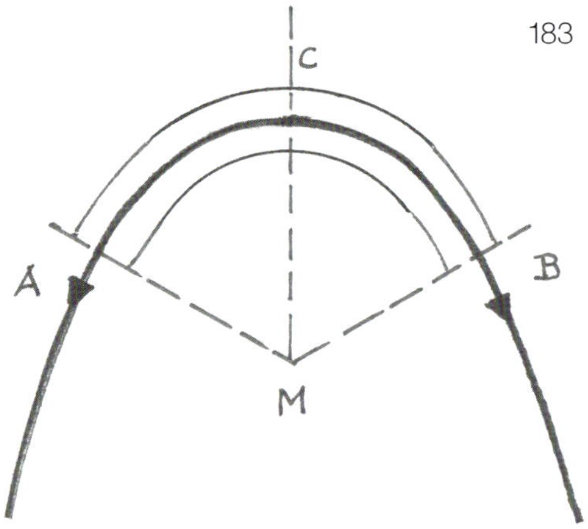

183

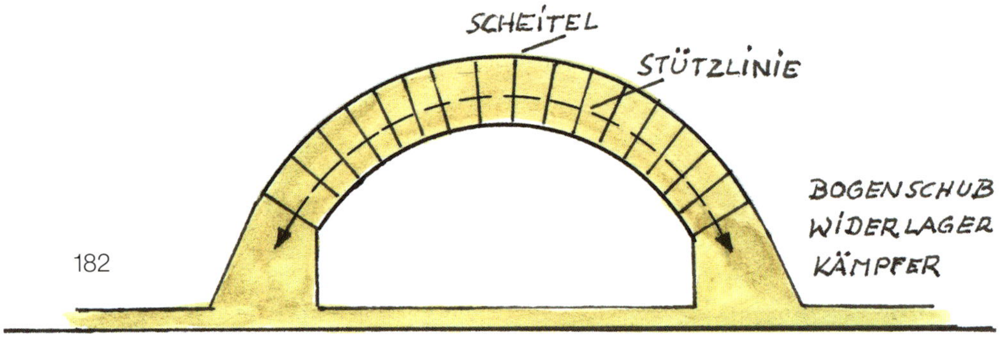

182

184, 185. Bridge over the Marsyas in southwest Turkey. Built with variable arch spans on bedrock, it was an aqueduct with pedestrian crossing.

5.7. Foundations

5.7.1. Soil conditions

Foundations shall provide a stable support of the weight of structures. Arches and vaults are highly sensitive to the lack of rigid support, in particular the settlement of foundations. Settlement will occur when the bottom of excavation has been disturbed (loosened), or when the load to be placed exceeds the weight of excavation. It is very difficult to handle differing ground conditions, such as bedrock on one end and soil on the other end of an arch. Thorough soil exploration is a must before construction. The only means available to the Romans was the digging of test pits or the use of iron sounding rods, which were driven into the ground. Their resistance indicated the soil quality. Whenever possible, the Romans tried to reach solid rock.

To improve soil conditions the Romans used manual compaction or admixtures of lime. The most effective method to improve bearing capacity of poor soils was piling: logs were driven into the ground at close spacing, thereby consolidating the soil.

Alberti (*De re aedificatoria*, 1452) prescribed rule-of-thumb formulae for the design of arched stone bridges: »The voussoirs of an arch should have a width of at least one tenth of the span.« The span of the arch should not be more than six or less than four times the widths of the piers. Vitruvius (VI 8) already stressed that the quality of foundations for vaulting is of utmost importance: »But if underground rooms and vaults are intended, their foundation ought to be thicker… In houses where piers are used in the construction, when there are arches composed of voussoirs with joints radiating to the center, the outermost piers must be made broader then the others, so they may have the strength to resist, when the wedges under the pressure of the load above begin to press along their joints out to the abutments.«

An advantage was the slow progress of construction in Roman times, allowing the ground to settle, before the superstructure was built.

5.7.2. Spread foundations

Spread foundations are used on good soils, the size of the footing being a function of the load and the bearing capacity of the ground. The allowable pressure from dead weight and live load ranged from 1 to 5 kg/cm^2. Calculation of a 12 m bridge indicated 5 kg/cm^2. On bedrock these values could be much increased, provided that a perfect fit of the masonry was achieved.

Dry summer periods were the best time to build in a riverbed. Otherwise the river had to be deviated as Herodot relates from a bridge at Babylon (H. I, 186). If ground conditions varied, the design of the bridge was adjusted, sometimes resulting in differing spans or changes in direction.

If work was necessary below the water level sheet piling was employed. The work area was surrounded by two walls of boarding with an infill of sealing clay, creating a cofferdam. After pumping out the water, excavation could begin and the foundation be laid. Vitruvius explains this technique in detail (V 12, 2–5): »Walls which are to be under water should be constructed as follows. Take the powder which comes from the country extending from Cumae to the promontory of Minerva (pozzuolana earth) and mix it with lime two to one. Then drive a cofferdam, with its sides formed of oaken stakes with ties between them, down into the water and prop it firmly on the ground. … Finally, concrete from the mortar trough must be heaped up until the empty space which was within the cofferdam is filled up by the wall.

… But in places where this powder is not found, the following method must be employed. A cofferdam composed of charred stakes fastened together with ties should be constructed, and clay in wicker baskets made of swamp rushes should be packed between the propos. After this has been filled in as closely as possible, set up your water screws, wheels and drums and let the space now bounded by the enclosure be emptied and dried.«

He then prescribes further digging until good bearing ground is reached, otherwise the bottom must be stakes with piles made of charred alder or olive wood, upon which the masonry can be placed.

The Enns bridge at Lauriacum (Enns, Austria) is of interest. Since it crosses a fast flowing river out of the Alps, the piers had to be especially protected against wash-out under the pointed front, or kolking on the rear. Therefore the stone foundations were surrounded by closely driven tree trunks of 20 to 30 cm thickness.

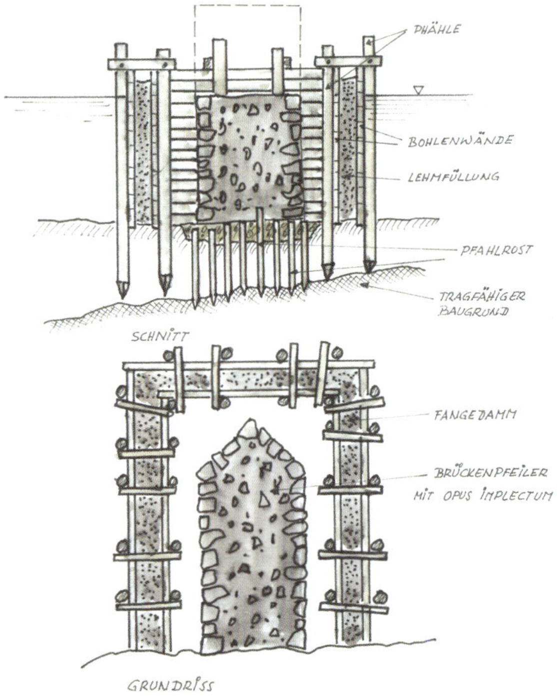

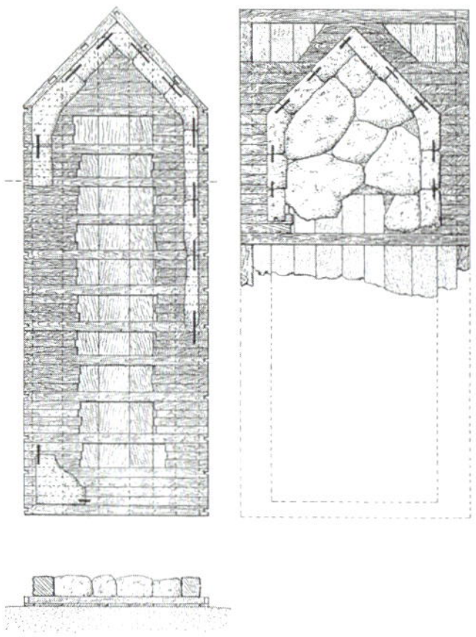

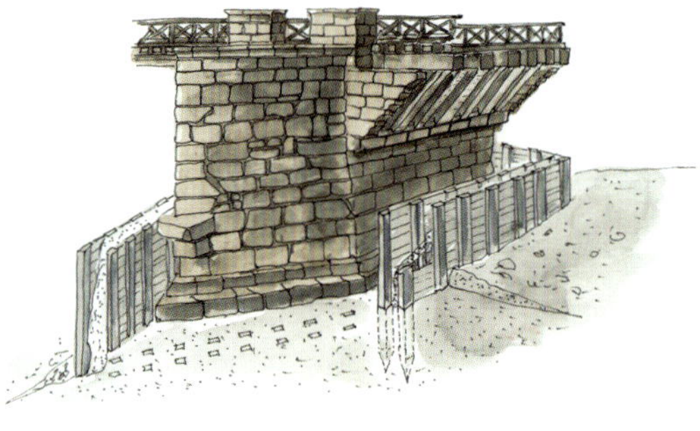

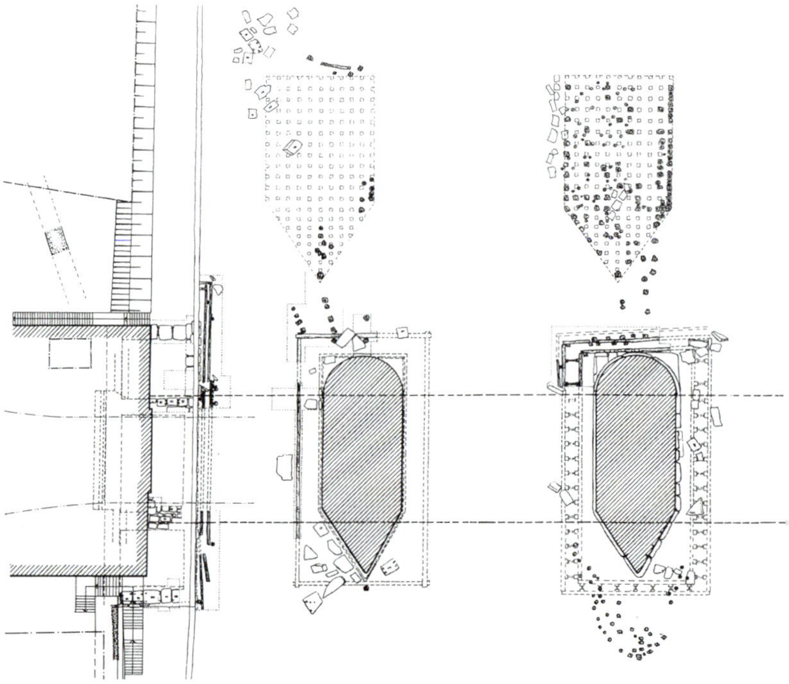

186. Section and plan of a cofferdam for the construction of a pier on piles. (Pfähle = piles, Bohlenwände = boarding, Lehmfüllung = clay infill, Pfahlrost = piling, tragfähiger Baugrund = solid ground, Schnitt = section, Fangdamm = cofferdam, Brückenpfeiler = rubble concrete pier, Grundriss = plan).
187. Cofferdam construction for a pier of the bridge over the Mosel at Trier (Cüppers 1969).
188. Pier foundation of the bridge over the Save River at Emona (Serbia). It had 16 piers at 12.30 m centers. A grid of wooden beams produced a kind of box, 9.7 x 3.45 m, pointed against the flow. It was filled with rubble and since wood does not decay if permanently submerged in water, this was a lasting solution (Cüppers, 1969).
189. Alcántara Bridge over the Tejo, built on bedrock.
190. Foundations of the first and second bridge over the Mosel at Trier (Cüppers 1969).

Case study: stone bridge in Trier

On the occasion of channeling works for the Mosel River in the years 1957 to 1967, entailing a deviation of the water, the *Rheinische Landesmuseum* was able to conduct detailed studies of Roman bridge foundations. H. Cüppers (1969) published the findings.

The stone bridge has shallow spread footings. Originally having nine piers plus the abutments, today seven piers remain. The river has an 1 to 3 m layer of gravel at the bottom, with clayey sandstone underneath. Bedrock lies several meters below the riverbed, necessitating about 5 m deep caissons of 26 by 12 m exterior dimensions. It is significant that there was a variety of solutions, due to different ground conditions and some experimentation.

For the deep version the walls consisted of two sets of horizontal oak wood logs, connected with braces and ties, creating two boxes, made water tight as far as possible. The void was filled with impervious material, such as clay. The inner working space for excavation was 24 by 10 m, having cross bracing to withstand the increasing external pressure as the caisson was slowly lowered, till it reached water tight ground.

For shallower situations simple wood piling was used, consisting of foot thick square piles, driven into the ground at 1.6 m centers. The piles had grooves to receive 7 cm thick horizontal boards. As variation vertical 20 x 25 cm wooden piles were driven next to each other and connected with horizontal beams.

An earlier bridge over the Mosel at Trier had seven piers with timber grates of 10 x 20 m as foundations. The piers themselves were 8 m wide and 10 m long plus 6 m at the foot for the pointed water break. Pile length ranged from 1.15 to 2.7 m, there were hundreds of them, with a round cross section, or four, six or eight sided, each 23 to 45 cm thick. On top of the vertical piles below water level an 80 cm layer of rip-rap was placed in cement, acting as a transition slab to the stone work.

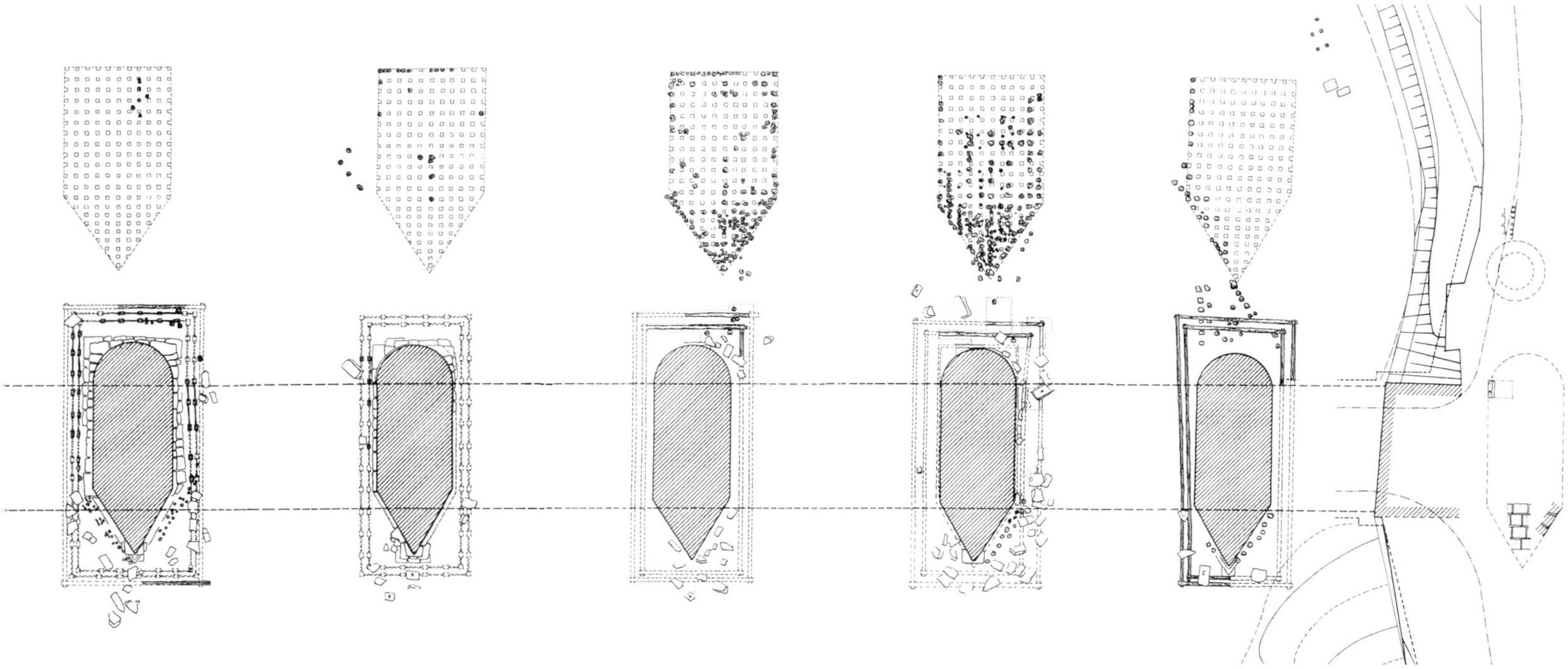

5.7.3. Deep foundations

When the bearing capacity of the riverbed is poor, it is more economical to drive supporting piles into the ground. It was very advantageous to work from above the water level, be it from a platform or from pontoons. Furthermore, driven piles improve the ground density by compacting it and the increasing resistance to each set of blows is an excellent indication of their carrying capacity. The strength of a pile derives from its contact with the ground at the bottom and the friction of the pile skin with the soil. Each round or squared wooden pile, about 6 m long and up to one foot thick was capped by a conic metal shoe, to facilitate driving. The pile grate was stopped below the water line to preserve the timber and tied together with a timber frame. Above it started the masonry pier.

Since rivers north of the Alps never went dry deep foundations were the rule. Given the abundance of good timber, many bridges were designed as timber frames on piles. Examples are the Rhine bridges at Mainz-Kassel (Cüppers 1969), at Koblenz (Schiefendecker 1980) and at Cologne (ditto).

191–193. Pile grate of the Rhine bridge at Mainz-Kastel (Cüppers, 1969).
194. Pier foundation of the Rhine bridge at Koblenz (Schieferdecker, 1980).
195. View of pile frame construction and pile driving operation (Köthe, 1991).
196. Front-view of piers on pile grates with coffer-dams (Heinz, 1989).
197. Side view of pier on pile grate, water flow from left to right.
198. Cross section of Caesar's bridge over the Rhine (Köthe, 1991).

191

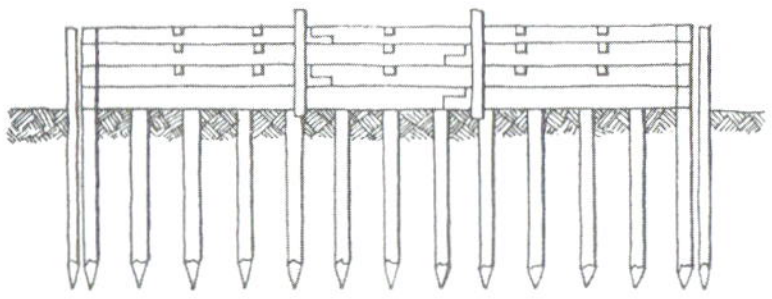

192

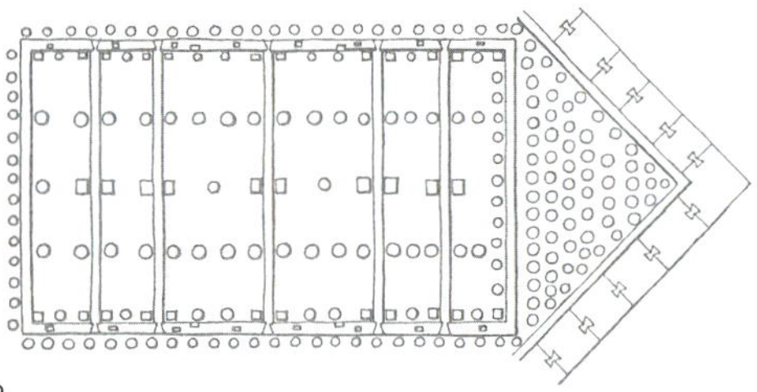

193

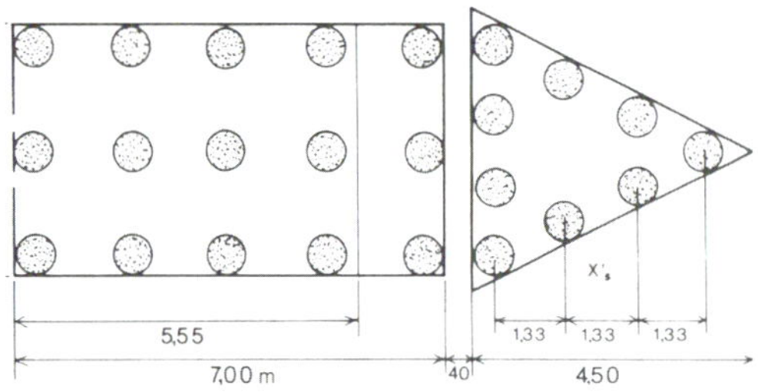

194

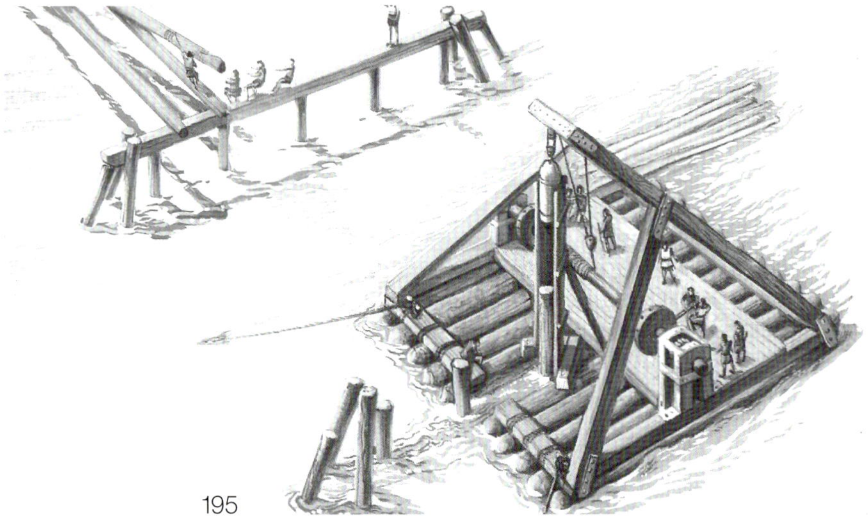

195

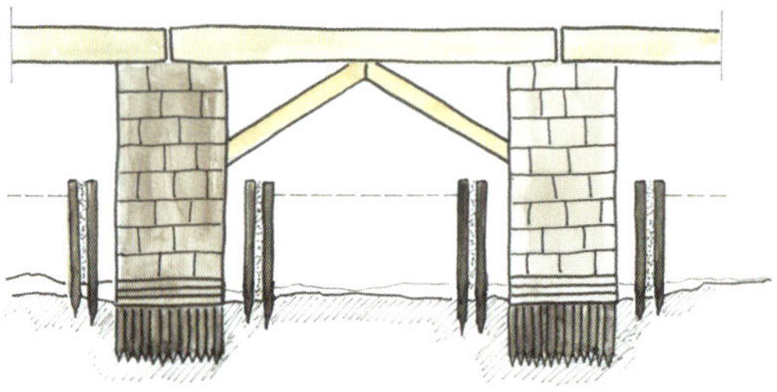

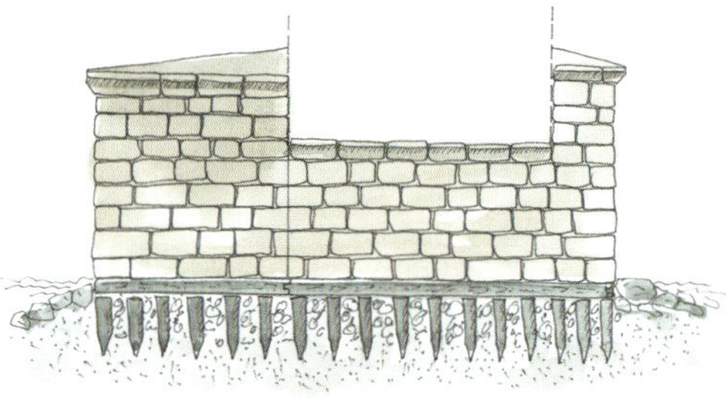

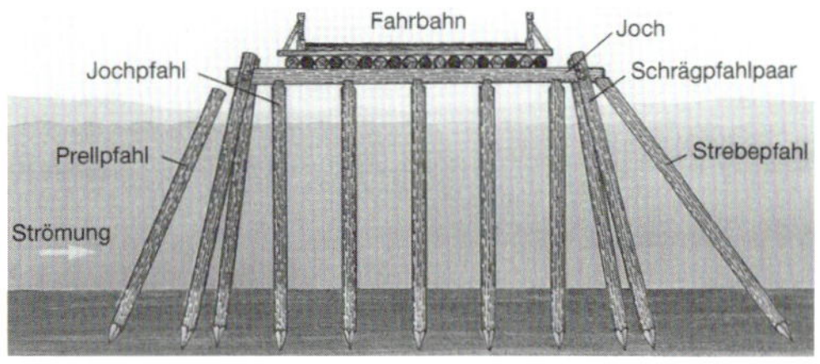

199. Distribution of forces between pier and abutment.
200. Superelevated middle arch of the Pont Julien, Vaucluse, France.

5.8. Piers

Piers are necessary when the bridge consists of more than one arch. Each pier has to absorb vertical and horizontal forces, resulting mainly from the dead weight of the structure. The horizontal forces of adjacent arches balance themselves and create a happy equilibrium. Service loads hardly amount to additional 10 % of the dead load and can be neglected, even by loading individual arches. However, if one arch fails, all the others are in danger of collapse. The abutments receive the resultant force from vertical load and horizontal thrust.

5.8.1. Dimensioning of piers

The height of a pier depended upon the cross section of the valley and the level of connecting roads. The tallest piers came close to 50 m.

The depth of a pier resulted from the width of the road, varying from 3 to 15 m. Adding the projecting cutwater in front and the rounding out at the back against kolking, the depth of a pier could reach 25 m.

The width of a pier depended upon the dead and service loads, as well as the height of voussoirs, since two of them had to have room side by side. Very rarely slender piers were in danger of buckling.

However, the width of the piers reduced the net width of the river and increased its speed, with all the negative consequences. The Roman builders had to reach a compromise. The 720 m long bridge at Mérida over the Guardiana has a flow width of 460 m, which is 64 %. Generally flow widths were around 70 %.

Flow width to full river width ratios of a few bridges:

62 %	Bridge in Cordoba, Spain
64 %	Puente Romano in Mérida, Spain
70 %	Ponte Milvio in Rome, Italy
70 %	Albarregas bridge in Mérida, Spain
73 %	Ponte d'Augusto in Rimini, Italy
74 %	Grande Ponte di Nona near Rome, Italy
75 %	Younger Mosel bridge at Trier, Germany
80 %	Ponte di Savignano, Forli, Italy
80 %	Bridge in Salamanca, Spain

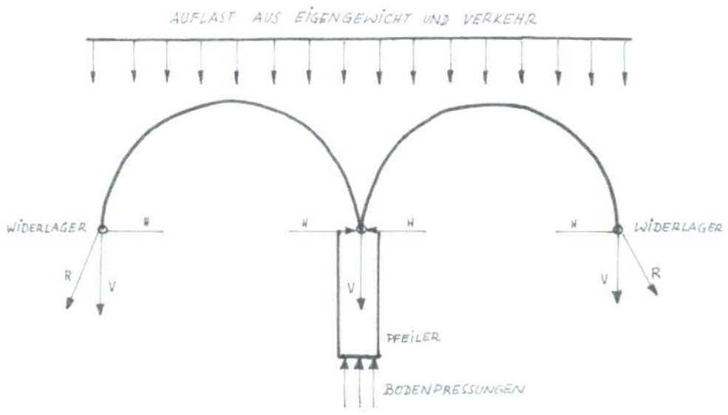

To increase the flow capacity in times of high water level, the piers sometimes had an opening at or above the impost level of the arches, called a floodway. In cases of wide riverbeds, which flooded rarely while the main flow kept to the middle, variable arch sizes could be used, with a major span in the center. This affords also passage of large boats and reduces the height of the piers if the approach sections are inclined. Often used in Asia Minor, the »hump« gives the bridge an oriental touch and probably offended the Roman sense of disciplined straightness.

When the moving water hits a pier, a horizontal water roll is generated. It is deflected to the sides, turns vertical and joins at the back of the pier, creating a whirlpool (kolking) which easily undermines the structure. Therefore, the lower upstream part of a pier had a pointed projection, deflecting the water like the prow of a ship and protecting the pier against flotsam or ice. On the downstream side a rounded projection eased the turbulence. Alberti (1485) considered the water behind a pier more dangerous than in front. Even with modern fluid mechanics formulae it is difficult to decide on the optimal shape against kolking. In 1976 the Reichsbrücke, the only bridge

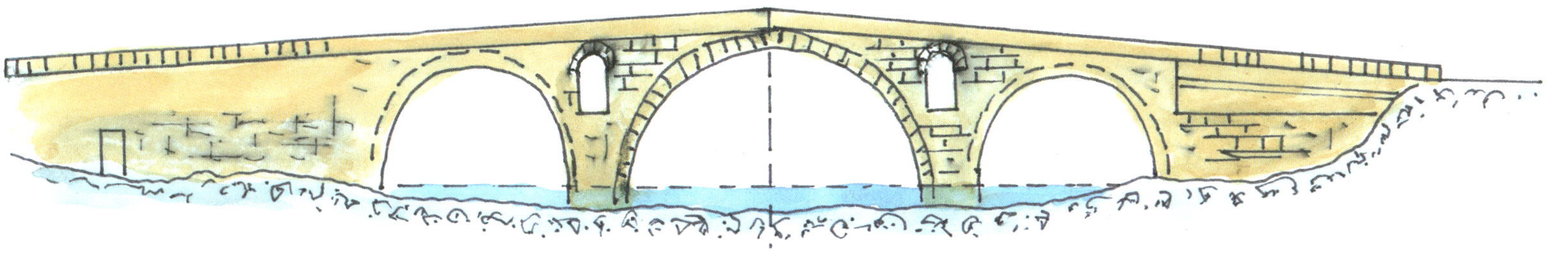

over the Danube in Vienna which survived the war, collapsed suddenly. It was a steel suspension bridge built in 1937. The disaster was largely attributed to kolking around the pylon foundations. Fortunately it happened at 5 a.m. and cost only one life.

Piers were very carefully built in ashlar masonry, the horizontal joints receiving a coat of lime to act as lubricant, which resulted in a near homogenous fabric. Mortar joints were avoided, they usually are the weak part of masonry. Generally the ashlars were perfectly dressed along the edges, slightly hollowed on the interior sides to give a perfect fit, while the exposed part of the blocks kept the quarry face. Masonry was laid in proper bond, with runners and headers sometimes further reinforced with iron cramps. The core consisted often of rubble cast in concrete, reducing the outer stones to a facing. Corbels were left projecting, to serve as support for falsework. In addition, continuous projecting moldings had the architectural effect of defining the springing of the arch.

201, 202. Flow conditions around a pier. Prallströmung = water roll, Auskolkung = material removed, Anlandung = material deposited, Spiralströmung = whirl flow, Ablösungsströmung = suction.
203. Bridge piers at Sommières, France, with cutwaters and floodways.
204. Detail of pier with cutwater (right) and protection against wash-out (kolking).
205. Simple cutwater pier of the bridge in Selinus (today Bergama, Turkey).

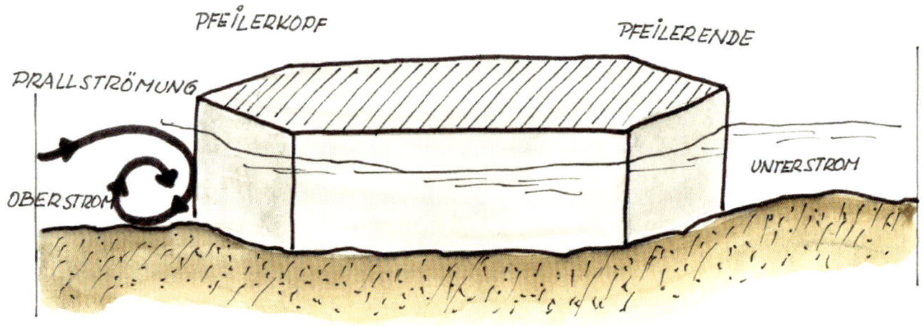

206–210. Arches with consoles to support false-work.
206. Bridge at Gualdo Tadino, Italy.
207. Bridge in Pergamon, Turkey.
208. Pont du Gard, France.
209. Augustus' Bridge at Narni, Italy. Note lower springing of left-side arch.
210. Continuous molding at the springing of the arches. Elja bridge at Segura, Portugal.

5.9. Abutments

The criteria regarding the size and construction of piers apply as well to abutments. Also here the dimensions depend on the topography of the location, the slope of the road and the width of the pavement (see 5.8 piers).

The abutments form the ends of the bridge on both sides. As the name indicates: That on which something abuts or is adjacent to; a part for sustaining or resisting pressure. It has to receive and to resist the forces of the vaulting.

As with piers, horizontal wooden constructions normally produce only vertical forces at the abutment. Horizontal forces from wind or braking vehicles can be neglected, given the great dead weight of Roman bridges. A vaulted structure with radial voussoirs however introduces significant horizontal forces, which do not cancel each other as they do on piers. The abutments are the most sensitive elements of a vaulting. Their rigidity is a precondition for the stability of a bridge. Does an abutment »give«, the adjacent arch will sag. When it collapses all further arches and piers will collapse in a domino effect, since the cancelling effect of the horizontal thrust of opposite arches is no more given.

The harmonious interplay of abutment, vaulting, piers and superstructure can be very well seen at the Pons Aelius, or Angels' Bridge in Rome.

If the ground at impost level is solid rock, the vault can be directly placed upon the prepared abutment surface. Solid rock is strong enough to absorb the horizontal forces. There is little danger of movement if the abutment is at the level of the general bridge foundation on bed rock. The high friction due to the great dead weight can be further increased by stepping of the foundation's base.

To secure also the inner mass of the vaulting against shifting, the core of the abutment within the outer stone casing should be a strong masonry or *opus caementitium.*

It becomes difficult to handle horizontal forces when the ground is not of solid rock but of a softer material, which sometimes may require piling. In this case the impost has to receive sufficient vertical forces to keep the resultant of vertical and horizontal forces within the core of the horizontal abutment's cross section. This way we avoid that, similar to piers, tensile forces generate at the edge of the abutment section. This is achieved by appropriate loads on top of the abutments. Since the circular arch produces automatically a higher road level, such additional loads are generally possible.

The bigger the span of a semicircular arch and its radius, the higher the road level. This increases the height of the abutment and with it the height of loading above the imposts, which in turn allows larger spans and the absorption of bigger horizontal forces.

The geometric forms and technical requirements from the properties of materials and the flow of forces complement each other. This is a further advantage of arched bridges using voussoirs.

The bearing capacity of the ground limits the additional loading of abutments. If it is exceeded it will result in dreaded settlements and failure of the arch. The only remedy would be piling.

When possible, bridges of a single arch were built at the narrowest locations of a valley. To keep the span small, the abutments were pushed towards the river, in which case wing walls were added, to direct the water past the abutments and to avoid erosion. The same applied when a bridge began in the middle of a river plain, being reached by a dam. At low water level the reduction of the flow capacity was no problem, but in high water wing walls had to protect the dams. The cost of dams was generally less than that of bridges.

5.10. Finishing

Along the sides of a bridge massive parapets with extra long blocks were built up to 80 cm high. They could be articulated by regularly projecting posts, giving a measure of scale. The architectural treatment would consist of projecting string courses and cornices, to articulate the different parts of the structure. Most of the parapets are gone, since the stones were easily removed and used otherwise.

Rustication gave a bridge an extra sense of durability. It could be applied as rock-faced, quarry-faced, smoothed or diamond pointed. Flat niches would receive commemorative panels and dedications. Normal road paving was continued over the bridge, generally irregular, polygonal or squared stone slabs in mortar. Rain water was drained to water spouts in the parapets. If the paving was not properly maintained it could lead to a deterioration of the vaulting. The Pont Flavius at Saint Chamas (Bouches-du-Rhône, France) is worn down from 1 m to 40 cm thickness.

211. Abutment and wing wall, Gualdo Tadino Bridge, Umbria, Italy.
212. Detail of abutment, Tejo bridge near Alconéta, Spain.
213. Basalt paving of the Alcántara Bridge and commemorative arch, Spain.
214. Worn-down pavement of the Pont Flavius at Saint Chamas, France.

6. Listing of bridges

(In the list case studies are indicated in bold numbers)

Northern Italy (IN)

	Via publica	Name, location	River	Province
IN 01		Ponte San Lorenzo, Padova		Padova
IN 02		Ponte Corbo, Padova	Bacchiglioni	Padova
IN 03		Ponte San Matteo, Padova		Padova
IN 04		Ponte Altinate, Padova	Brenta	Padova
IN 05		Ponte degli Angeli, Vicenza		Vicenza
IN 06	Postumia	Ponte Furo, Vicenza		Vicenza
IN 07	Postumia	Ponte Pusterla, Vicenza	Bacchiglioni	Vicenza
IN 08	Claudia Augusta	Ponte di Susegana, Susegana		Treviso
IN 09	Annia	Ponte sul Lemene, Concordia Sigittaria		Venezia
IN 10	Annia	Ponte di Fiumicinetto, Cittanova	Fiumecinetto	Venezia
IN 11	Annia	Ponte sul Canalat, Ceggia	Canalat	Venezia
IN 12	Annia	Ponte sul Natissa, Monastero di Aquleia Natissa		Friuli
IN 13	Claudia Augusta	Three bridges at Colfoso		Friuli
IN 14		Ponte Sonli, Sonli	Isonzo	Friuli
IN 15		Ponte Pietra, Verona	Adige	Verona
IN 16	Aemilia	Ponte sul Rubicon, Savignano	Rubicon	Forli
IN 17	Aemilia	Ponte San Vito, Rimini	Uso	Forli
IN 18	Aemilia	Ponte sul Torrente Uso, Sant'Arcangelo	Uso	Forli
IN 19	Aemilia	Ponte d'Augusto, Rimini	Marecchia	Forli
IN 20	Aemilia	Ponte sul Reno, Bologna	Reno	Bologna
IN 21	Aemilia	Ponte Petreo, Ravenna	Fossa Augusta	Forli
IN 22	Aemilia	Eight bridges at Parma, Parma		Parma
IN 23		Ponte della Regina, Almenno	Brembo	Bergamo
IN 24		Ponte di Corso Vittorio Emanuele, Milan		Milano
IN 25		Ponte de Arena, Monza		Bergamo
IN 26		Ponte Marzio, Gorle		Bergamo
IN 27		Ponte sull'Adda, Oliginate	Adda	Como
IN 28		Ponte Secchia plus two more		Emilia-Romana, Modena
IN 29		Ponte in Val Cannobio, Lago Maggiore		Varese
IN 30		Ponte di Promontogno Valtellina		Sondrio
IN 31		Ponte sulla Dora, Ivrea	Dora	Torino
IN 32		Ponte Ladricio, Fortezza	Isarco	Bolzano
IN 33		Ponte sul Passiro, Merano	Passei	Bolzano
IN 34		Ponte di Lagundo, Merano	Adige	Bolzano
IN 35		Pont St. Martin, San Martino	Lys	Aosta
IN 36		Ponte del Pondel, Val di Cogne		Aosta
IN 37		Ponte sul Torrente Buthier, Aosta		Buthier, Aosta
IN 38		Ponte sul Torrente Cillau, Saint Vicent		Cillau, Aosta
IN 39		Ponte sul Torrente Marmore, Chatillon	Marmore	Aosta
IN 40		Pont Bozet, Val Champorcher		Aosta
IN 41	Julia Augusta	Second bridge in Val Quazzola,		
IN 42	Julia Augusta	Four bridges in Val Quazzola, Quiliano		Savona
IN 43	Julia Augusta	Ponte di Val Ponci, Finale		Savona

	Via publica	Name, location	River	Province
IN 44	Julia Augusta	Four bridges in Val Ponci, Finale		Savona
IN 45	Julia Augusta	Pontetto, Luano		Savona
IN 46	Julia Augusta	Pontaccio, Luano		Savona
IN 47	Julia Augusta	Ponte sul Rio della Torre, San Stefano al Mare	Torre	Savona
IN 48	Julia Augusta	Ponte Lungo, Albenga		Savona
IN 49		Ponte Vecchio, Rapallo		Genova
IN 50		Ponte dell'Arsina, Arsina	Vara	La Spezia

Southern Italy (IS)

	Via publica	Name, location	River	Province
IS 01	Appia	Ponte di Minturno	Garigliano	Caserta
IS 02	Appia	Ponte Tufaro, Benevento		Benevento
IS 03	Appia	Ponte Apollosa, Benevento		Benevento
IS 04	Appia	Ponte Leproso, Benevento		Benevento
IS 05	Appia	Ponte Corvo, Benevento		Benevento
IS 06	Appia	Ponte del Volturno, Capua	Volturno	Caserta
IS 07	Appia	Ponte Pietra, Canosa di Puglia	Olfantu	Bari
IS 08	Appia	Ponte di Ladroni, Benenvento		Benevento
IS 09	Appia	Ponte dell Chianche, Montecalvo		Benevento
IS 10	Appia	Pons Auruncus, Sessa Aurunca		Napoli
IS 11	Traiana	Ponticello, Benevento		Benevento
IS 12	Traiana	Ponte sul Carapelle, Herdoniae		Carapelle, Foggia
IS 13	Traiana	Ponte sul Cervaro, Herdoniae		Cervaro, Foggia
IS 14	Traiana	Viadotto di Apani, Apani		Brindisi
IS 15		Ponticello di Iserna, Iserna		Campobasso
IS 16		Ponte del Diavolo	Titerno	Campobasso
IS 17		Ponte sul Savuto, Scigliano	Savuto	Cosenza
IS 18		Ponte Romano, Amalfi		Salerno
IS 19		Ponte sul Olfanto, Santa Venere	Olfanto	Foggia
IS 20		Ponte Valentino, Pesto	Calore	Salerno
IS 21		Ponticello di Selinunt,	Selino	Sicily
IS 22		Ponte sul Manno, Porto Torres	Mannu	Sardinia
IS 23		Ponte di Sant'Anticio, Sant'Anticio		Sardinia
IS 24		Ponte di Gavoi, Gavoi		Sardinia
IS 25		Ponte di Tirso, Oristano	Tirso	Sardinia

Central Italy (IM)

	Via publica	Name, location	River	Province
IM 01	Flaminia	Milvius Pons, Rome	Tiber	Urbs Roma
IM 02		Fabricius Pons, Rome	Tiber	Urbs Roma
IM 03		Aemilius Pons, Rome	Tiber	Urbs Roma
IM 04		Aelius Pons, Rome	Tiber	Urbs Roma
IM 05		Cestius Pons, Rome	Tiber	Urbs Roma
IM 06	Tiburtina	Ponte Mammolo, Rome	Anio	Urbs Roma
IM 07		Fosso di Prima Porta, Rome		Urbs Roma
IM 08	Labicana	Ponte sul Fosso Giardinetta, Rome		Urbs Roma
IM 09		Lost bridges in Rome: Pons Aurelius, Agrippa, Neroianus, Sublicius, Probi		
IM 10	Appia	Viadotto di valle Ariccia, Ariccia		Roma
IM 11	Appia	Ponte Catena, Cori		Latina

	Via publica	Name, location	River	Province
IM 12	Appia	Ponte di Vigna Capoccio, Lanuvio	Fosso di Carano	Roma
IM 13	Appia	Ponte di tre ponte, Treponte	Ninfa	Latina
IM 14	Appia	Ponte sul Cavala, Faiti		Latina
IM 15	Appia	Ponte Alto, Terracina		Latina
IM 16	Appia	Ponte Antico (P. Granci), Terracina		Latina
IM 17	Flaminia	Ponte Torre Pastore, Civita Castellana		Viterbo
IM 18	Flaminia	Muro del Peccato, Civita Castellana	Treia	Viterbo
IM 19	Flaminia	Viadotto della Treia, Civita Castellana	Treia	Viterbo
IM 20	Flaminia	Bridge in Civita Catellana	Treia	Viterbo
IM 21	Flaminia	Fosso della Valchetta	Veio	Roma
IM 22	Flaminia	Ponte Picchiato, Orte	Miccino	Viterbo
IM 23	Flaminia	Ponte Ritorto, Civita Castellana	Fosso Chiavello	Viterbo
IM 24	Flaminia	Ponte Rio Franta, Borghetto	Fratta	Viterbo
IM 25	Flaminia	Ponte Minucio near Otricoli ?	Tiber	Viterbo
IM 26	Flaminia	Ponte Sanguiario, Otricoli		Terni
IM 27	Flaminia	Ponte di Augusto, Narni	Nerva	Terni
IM 28	Flaminia	Ponte Calamone, Terni		Terni
IM 29	Flaminia	Ponte Cardaro	Terni	Terni
IM 30	Flaminia	Ponte sul Velino, Villa di Valle	Velino	Terni
IM 31	Flaminia	Ponte Toro	Papigno	Terni
IM 32	Flaminia	Ponte Fonnay, Massa Martana	Bevegna	Perugia
IM 33	Flaminia	Ponte di Giovanni de Butris, Massa Martana		Perugia
IM 34	Flaminia	Ponte del Diavolo, Massa Martana	Bevegna	Perugia
IM 35	Flaminia	Primo Ponte in Foligno, Foligno		Perugia
IM 36		Secundo Ponte in Foligno, Foligno		Perugia
IM 37		Terzo Ponte in Foligno, Foligno		Perugia
IM 38		Quarto Ponte in Foligno, Foligno		Perugia
IM 39	Flaminia	Ponte Centesimo, Foligno	Topino	Perugia
IM 40	Flaminia	Ponte Marmoreo, Nocera Umbra		Perugia
IM 41	Flaminia	Ponte Sanguinario, Spoleto	Tessino	Perugia
IM 42	Flaminia	Gualdo Tadino, Primo Ponte, Sigillo		Umbria
IM 43	Flaminia	Ponte sul Fosso Regalino, Fossato di Vico		Perugia
IM 44	Flaminia	Ponte San Giovanni, Fossato di Vico		Perugia
IM 45	Flaminia	Ponte di Pioraco, Pioraco	Scheggia	Macerata
IM 46	Flaminia	Ponte Grosso sul Fosso della Scheggia, Pontericcioli	Scheggia	Perugia
IM 47	Flaminia	Ponte Voragine, Pontericcioli	Furicchie	Perugia
IM 48	Flaminia	Ponte detto Etrusco, Costacciaro		Perugia
IM 49	Flaminia	Ponte Grosso, Cagli	Burano	Perugia
IM 50	Flaminia	Ponte Manlio, Cagli	Bosso	Pesaro
IM 51	Flaminia	Ponte di Smirra, Smirra		Pesaro
IM 52	Flaminia	Ponte Taverna, Cagli	Burano	Pesaro
IM 53	Flaminia	Ponte die Ciclopi, Cagli		Pesaro
IM 54	Flaminia	Ponticello presso il Ponte dei Ciclopi, Cagli		Pesaro
IM 55	Flaminia	Ponticello di Casenuove, Casenuove		Pesaro
IM 56	Flaminia	Ponticello dell'Abbazia, Acqualonga		Pesaro
IM 57	Flaminia	Ponte di Diocleziano, Fossombrone		Pesaro
IM 58	Flaminia	Ponte di Traiano, Fossombrone	Metauro	Pesaro

	Via publica	Name, location	River	Province
IM 59	Flaminia	Ponte do Porta Rimini, Pesaro	Fóglia	Pesaro
IM 60	Flaminia	Ponte di Pollenza, Pollenza		Mercata
IM 61	Flaminia	Ponte Spanio, Spanio-Sigillo	Fonturce	Ancona
IM 62	Flaminia	Ponte del Piangelo, Fano	Metauro	Ascoli
IM 63		Ponte sul Torrente Lorenzino, Caudino		Ancona
IM 64	Ostiense	Fossa Malafede, Torre Pastore		Roma
IM 65	Ostiense	Ponte di Tor di Valli		Roma
IM 66	Ostiense	Ponte di Mezzocamino		Roma
IM 67	Aurelia	Viadotto sulla Via Aurelia, Rome		Roma
IM 68	Aurelia	First bridge, Santa Marinella	Fosso Morgana	Roma
IM 69	Aurelia	Second bridge, Santa Marinella	Fosso Morgana	Roma
IM 70	Aurelia	Ponte Romano, Santa Marinella Fosso di Valemplice		Roma
IM 71	Aurelia	Ponte del Diavolo, Santa Marinella	Fiumretta	Roma
IM 72	Praenestina	Piccolo Ponte, Nona		Roma
IM 73	Praenestina	Grande Ponte, Nona		Roma
IM 74	Praenestina	Ponte Amato, Gallicano	Fosso dell'Infernacio	Roma
IM 75	Praenestina	Ponte Sardone, Sardone,		Roma
IM 76	Praenestina	Pons Gabii, Gabii	Fosso d'Acqua Rossa	Roma
IM 77	Cassia	Ponte Sodo, Veio near Isola Farnese		Roma
IM 78	Cassia	Ponte Camarillo, Bulicane		Viterbo
IM 79	Cassia	Ponte San Nicola, Sette Vene		Viterbo
IM 80	Cassia	Ponte di Bolsena, Bolsena	Fosso Melona	Viterbo
IM 81	Clodia	Ponte del Diavolo, Manziana		Roma
IM 82	Clodia	Ponte della Rocca, Bieda		Viterbo
IM 83	Clodia	Ponte del Diavolo, Bieda		Viterbo
IM 84	Clodia	Ponte Piro, Barbarano		Viterbo
IM 85	Valeria	Ponte Funicchio, Ferento		Viterbo
IM 86	Valeria	Ponte San Giorgio, Arsoli		Roma
IM 87	Valeria	Ponte Scutonico, Arsoli		Roma
IM 88	Tiburtina	Ponte Marmone, Arpino	Liris	Frosinone
IM 89	Tiburtina	Ponte Lucano, Tivoli	Anio	Roma
IM 90	Tiburtina	Ponte dell'Acquoria	Tivoli	Roma
IM 91	Salaria	Ponte Sambuco, Rieti	-:-	Rieti
IM 92	Salaria	Ponte Salario, Roma	Anio	Roma
IM 93	Salaria	Ponte sul Fiume Velino, Rieti	Velino	Rieti
IM 94	Salaria	Ponte Mercato, Acquaviva		Rieti
IM 95	Salaria	Ponte Bruciato, Buida		Rieti
IM 96	Salaria	Ponte di Porta Cappucina, Ascoli Piceno	Tronto	Ascoli
IM 97	Salaria	Ponte di Cecco, Ascoli Piceno	Tronto	Ascoli
IM 98	Salaria	Ponte sul Garafo, Acquasanta	Garafo	Ascoli
IM 99	Salaria	Ponte della Scutella, Scutella	Cavignano	Ascoli
IM 100	Salaria	Ponte del Gran Caso, Ascoli		Ascoli
IM 101	Salaria	Ponte Quindodecimo, Quindodecimo		Ascoli
IM 102	Salaria	Ponte d'Arli, Arli	Tronto	Ascoli
IM 103	Salaria	Ponte do Porta Tufilla, Ascoli Piceno	Tronto	Ascoli
IM 104	Salaria	Ponte Mozzana, Mozzana	Tronto	Ascoli
IM 105		Ponte sul Tevere, Sigliano	Tiber	Arezzo
IM 106		Ponte sul Sentino, Genga	Sentino	Pesaro
IM 107		Pons Nomentano, Roma	Anio	Roma

	Name, location	River	Province
IM 108	Ponte sul Fosso Tre Ponti, Civita Castellana		Viterbo
IM 109	Ponte di Nepi, Civita Casrellana		Viterbo
IM 110	Ponte del Diavolo, Pignataro	Apalla Bassa	Frosinone
IM 111	Ponte nella Valle des Rapido, Cassino	Rapido	Frosinone
IM 112	Ponte San Elia Vecchio, Cassino		Frosinone
IM 113	Ponte Nascoso, Civitatomassa		Abruzzi
IM 114	Ponte di Sperlonga, Sperlonga		Latina
IM 115	Ponte San Pietro, Poli		Roma
IM 116	Ponte di San Antonio, Gallicano	Acqua Raminga	Roma
IM 117	Ponte Lupo (aqueduct), Poli		Roma
IM 118	Ponte sul Fosso Caipoli, Poli	Fosso Caipoli	Roma
IM 119	Ponte della Bulica (aqueduct), Poli		Roma
IM 120	Bridge at Rota, between Civitavecchia and Bracciano		Roma
IM 121	Ponte sul Mar, Gravisca		Roma
IM 122	Ponte sul Fosso di Terra, San Vittorin		Roma
IM 123	Ponte sull'Aniene, Agosta	Anio	Roma
IM 124	Ponte di Fontana Nuova, Sacrofano		Latina
IM 125	Arci San Vidam, between Sermoneta and Latina Scala		Latina
IM 126	Ponte sul Fosso della Mola, Castel San Elia		Viterbo
IM 127	Ponte dell'Abbadia, Vulci	Fiora	Viterbo
IM 128	Ponte di San Giuliano, Atina	Rapido	Frosinone

France

	Name, location	River	Department
F 01	Pont Romain, Saint-Benoit-du-Sault	Le Portefeuille	Indre
F 02	Pont Romain, Les Clefs near Annecy	Le Fier	Haute-Savoie
F 03	Pont des Esclapes, Fréjus	Argens	Var
F 04	Pont d'Ambroix, Ambroix	Vidourle	Gard
F 05	Pont de Vendelais, Vendelais	Couesnon	Ille-et-Vilaine
F 06	Pont de St. Thibery, St. Thibery	Hérault	Hérault
F 07	Pont Serme, Serme	Orb	Hérault
F 08	Pont de Lodève, Sommières	Vidourle	Gard
F 09	Pont de Boissezon, Boissezon	Bénovie	Hérault
F 10	Pont Julien, Apt	Calavon	Vaucluse
F 11	Pont Romain, Vaison la Romaine	Ouvèze	Vaucluse
F 12	Pont de Mane, Mane	Laye	Provence-Alpes-Côtes d'Azur
F 13	Pont Flavien, Saint-Chamas		Bouches-du-Rhône
F 14	Pont Romain, Domqueur,		Somme
F 15	Pont Romain, Outremecourt	Mouzon	Haute-Marne
F 16	Pont Romain, Sainte-Colombe	Rhône	Doubs
F 17	Pont Romain, Ascain	Nivelle	Pyrénées-Atlantique
F 18	Pont Romain, Lavoute	Loire	Haute-Loire
F 19	Pont St. Gabriel, Tarascon		Bouches-du-Rhône

	Name, location	River	Department
F 20	Pont Romain, Sommières	Vidourle	Gard
F 21	Pont Romain, Arles	Rhône	Provence-Alpes-Côtes-d'Azur
F 22	Pont Romain, Rolampont	Marne	Haute-Marne
F 23	Pont Vieux, Le Vignan		Gard
F 24	Arc de Glanum, Saint-Rémy		Bouches-du-Rhône
F 25	Pont Romain, Vivier	Escoutay	Ardèche
F 26	Pont du Gard, Rémoulins	Le Gardon	Gard
F 27	Pont Vieux, Béziers	Orb	Hérault
F 28	Pont Romain, Ganagobie	Buès	Alpes-de-Haute-Provence
F 29	Pont Romain, Fugeret	La Va	Alpes-de-Haute-Provence
F 30	Pont Romain, Chantelle	Bouble	Allier
F 31	Pont Romain, Spoy	Landion	Champagne-Ardenne
F 32	Pont de Crau, Arles		Provence-Alpes-Côtes-d'Azur
F 33	Arch of Germanucus, Saintes	Charante	Charante-Maritime
F 34	Pont Romain, Mouthiers d'Ahun	La Creuse	Creuse
F 35	Pont d'Apas, Saint-Martory	La Garonne	Haute-Garonne / Midi-Pyrénées
F 36	Pont Romain, Chalon-sur-Saône		Saône-et-Loire
F 37	Pont Vétuste, Narbonne		Aude
F 38	Pont Romain, Saint-Christophe-la-Grotte		Savoire

Belgium (B), Luxembourg (L) , Andorra (A)

	Name, location	River	Country
B 01	Pont Montignies, Saint-Christoph	Hantes	Belgium
B 02	Pont d'Ombret, Arlon-la-Semois		Belgium
L 01	Pont Romain, Echternach	Saar	Luxembourg
L 02	Pont Romain, Wasserbillig	Saar	Luxembourg
A 01	Pont Romain	Valira	Andorra

Spain (S) and Portugal (P)

	Name, location	River	Province
SP 01	Puente de Piedra, Zaragossa	Ebro	Zaragossa (S)
SP 02	Salamanca Bridge, Salamanca	Tormes	Salamanca (S)
SP 03	Sella bridge, Cangas de Onis	Sella	Asturia (S)
SP 04	Fluvia bridge, Besalù	Fluvia	Gerona (S)
SP 05	Bridge at Zamora, Zamora		Tarragona (S)
SP 07	Bridge at Cordubar, Córdubar	Guadalquivir	Córduba (S)
SP 08	Salado bridge	Salado	Cordoba (S)
SP 09	Bridge at Mérida, Mérida	Guadiana	Bajadoz (S)
SP 10	Pedrina bridge, Limia, Bande		Orense (S)
SP 11	Bridge at Alconétar, Alconétar	Tejo	Caceres (S)
SP 12	Bridge at Mérida, Mérida	Albarregas	Bajadoz (S)
SP 13	Puente del Diablo, Martorell	Lobregat	Catalonia (S)

	Name, location	River	Province
SP 14	Puente del Monistrol, Monistrol	Lobregat	Càceres (S)
SP 15	Ambroz Bridge, Ambroz, Ventas de Caparra		Caceres (S)
SP 16	Trespuntes de Iruna	Zadorra	Alava (S)
SP 17	Bridge at Villodas, Villodas	Zadorra	Alava (S)
SP 18	Alcantarilla Bridge, Seville	Salado	Sevillia (S)
SP 19	Puebla de Trives Bridge, Puebla di Trives	Bibey	Orense (S)
SP 20	Bridge at Lugo, Lugo, Jiloca		Tarragona
SP 21	Bridge at Vadillo, Vadillo	Tamuja	Cárceres (S)
SP 22	Bridge at Mérida, Mérida	Brodalo	Bajadoz
SP 23	Bridge at El Garro, El Garro	Almonte	Estramadura (S)
SP 24	Bridge at La Vera, Al Vera	Alardos	Cáceres (S)
SP 25	Bridge at La Vera, Al Vera	Gullamino	Cáceres (S)
SP 26	Bridge at Doncella, Doncella	Ambroz	Cáceres (S)
SP 27	Bridge at Medellin, Medellin	Anas	Bajadoz (S)
SP 28	Alcántara Bridge, Alcántara	Tejo	Toledo (S)
SP 29	Trajan's Bridge, Alcántara	Tejo	Cáceres (S)
SP 30	Bridge at Guijo, Guijo de Granadilla	Alagón	Cáceres (S)
SP 31	Orbigo Bridge, Santiago de Compostela		Corûna (S)
SP 32	Bridge at Almunécar, Almunécar	Marchan	Costa del Sol (S)
SP 33	Bridge at Assa, Assa	Ebro	Rioja (S)
SP 34	Bridge at Caesar Augusta, Persola	Isuela	Pyrenees (S)
SP 35	Bridge at Caesar Augusta, Persola,	Huesac	Pyrenees (S)
SP 36	Bridge at Golmayo, Golmayo, Rioja		Soria (S)
SP 37	Bridge in Granada Province		Granada (S)
SP 38	Puente de Carmona, Carmona		Sevilla (S)
SP 39	Puente de la Cigarrosa		Ourense(S)
SP 40	Bibey bridge	Bibey	Spain
SP 41	Bridge at Chaves, Chaves	Tamega	Vila Real (district)
SP 42	Bridge at Alter do Chao, Villa Formosa	Seda	Ribatejo (P)
SP 43	Bridge at Segura, Segura	Elja	Alentejo (P)
SP 44	Bridge at Transmontana, Transmontana	Tuela	North Portugal
SP 45	Two bridges at Viseo, Viseo	Dad	Beira Alta (P)

Germany, Austria, Switzerland

	Name, location	River	Country
G 01	Mosel bridge, Trier	Mosel	Germany
G 02	Mosel bridge, Palzem-Stadtbredimus	Mosel	Germany
G 03	Mosel bridge, Koblenz	Mosel	Germany
G 04	Twenty-seven bridges along the Mosel River		Germany
G 05	Rhine bridge, Mainz-Kastel	Rhine	Germany
G 06	Rhine bridge, Cologne (Köln)	Rhine	Germany
G 08	Twelve bridges along the Rhine		Germany and Switzerland
G 09	Six bridges along the Neckar River		Germany
G 10	Ten bridges along the Main and the Jagst Rivers		Germany

	Name, location	River	Country
G 11	Ponte Aeni, Pfaffenhofen	Inn	Germany
G 12	Pont de Nudry, Grand St. Bernard		Switzerland
G 13	Bridge at Intragna, Ticino		Switzerland
G 14	St. Dionysus Bridge, St. Dionysen, Styria	Mur	Austria
G 15	Bridge at Oberzeiring, Blahbach near Graz, Styria		Austria
G 16	Nine bridges in Switzerland and Austria		

The Balkans

	Name, location	River	Country
B 01	Emona Bridge, Aquileia	Sava	Croatia
B 02	Five bridges at Salonae, Spalato		Croatia
B 03	Bridge at Nish, Nish	Nishava	Serbia
B 04	Six bridges in Serbia, Macedonia and Montenegro		
B 05	Bridge at Köstendil, Köstendil	Struma	Bulgaria
B 06	Danube bridge, Turnu Severin	Danube	Romania
B 07	Danube bridge, Czernecz	Danube	Romania
B 08	Bridge at Elbasan, Elbasan	Genusus (Shkumbi)	Albania
B 09	Nine bridges along the Via Egnetia over the Shenjan and Shkumbi		Albania
B 10	Bridge on Rhodos, Rhodos		Greece
B 11	Bridge at Vrousina, Vrousina	Thyamis (Kalamas)	Greece
B 12	Bridge at Eleusis, near Athens	Kephissos	Greece

Turkey

	Name, location	River	Province
T 01	Mouslouk Bridge, Bergama	Selinus	Izmir
T 02	Twin Tunnel, Bergama	Selinus	Pergamon
T 03	Lower Bridge, Bergama	Selinus	Pergamon
T 04	Fourth Bridge, Bergama	Selinus	Pergamon
T 05	Bridge at Gergas, Ince-Kemer	Marsias	Aydin
T 06	Bridge at Laodiceia, Eski Hisar	Licus	Denizli
T 07	Bridge at Azanoi, Azanoi	Rhyndacus	Kütahya
T 08	Bridge at Kütahya, Kütahya	Rhyndacus	Kütahya
T 09	Nysa Bridge, Aydin	Cakircak	Nysa
T 10	Aqueduct Bridge, Kemer Dere (Troy)		Cannakale
T 11	Ilica Dere Bridge, Ilica Dere, Troy		Cannakale
T 12	Justinian's Bridge, Adapazari	Sakaria	Pergamon
T 13	Bridge at Mysien, Mysien	Makestos	Balikesir
T 14	Bridge at Beskonak, Aspendos	Eurymedon	Antalya
T 15	Bridge at Beskonak off Eurymedon	Selge	Antalya
T 16	Bridge at Toruncova (Limyra)	Alagir Cay	Finike
T 17	Bridge at Antalya, Düden, Antalya		Antalya
T 18	Bridge at Antalya, Antalya		Antalya
T 19	Bridge at Aspendos, Aspendos	Eurymedin	Antalya
T 20	Bridge at Manavgat, Side	Manavgat Cay	Antalya
T 21	Two bridges, Telmessos		Antalya
T 22	Bridge at Kemer, Kemer	Xanthos	Antalya
T 23	Bridge at Antalya, Antalya	Arab Su	Antalya

	Name, location	River	Province
T 24	Bridge at Seleucia, Silifke		Antalya
T 25	Bridge at Tarsus, Tarsus		Mersin
T 26	Bridge at Adana, Adana	Seyhan	Adana
T 27	Bridge at Adana, Misis		Adana
T 28	Bridge at Antakya, Antakya	Asi	Hatay
T 29	Xanthos bridge, Seki	Xanthos (Kocacay)	Denizli
T 30	Three bridges, Eski Hisar, Milas		Antalya
T 31	Bridge at Olympos, Antalya		Antalya
T 32	Bridge at Aperlai, Kekova		Antalya
T 33	Bridges at Ankara, Ankara		Ankara
T 34	Bridge at Kakho, Kakho		East Turkey
T 35	Chabinas bridge, Eski Kahta	Chabinas	Adiyaman
T 36	Bridge at Killisik, Killisk	Karasu	East Turkey
T 37	Karas bridge	Karas	Erzurum
T 38	Bridge at Adiyaman, Adiyaman	Gök-Su	Adiyaman
T 39	Two bridges	Arabkir	SE Turkey
T 40	Karabudak Bridge	Euphrates	SE Turkey

North Africa

	Name, location	River	Country
A 01	Bridges at Theveste and Gastal, Tebessa		Algeria
A 02	Bridge at Biskra, El Kantara		Algeria
A 03	Bridge at Cirta, Constantine	Oued Rhumel	Algeria
A 04	Bridge at Aquedotto, Khamissa		Algeria
A 05	Pont de Trajan	Oued Beja	Tunisia
A 06	Bulla Regia Bridge, Chemtou	Bagradas	Tunisia
A 07	Bridge at el Oued, Grich el Oued	Bagradas	Tunisia
A 08	Bridge at Djedeida, Djedeida	Bagradas	Tunisia
A 10	Bridge at El Knatit, El Knatit	Oued Cherchar	Tunisia
A 11	Bridge at Duperré, Duperré	Chelif	Tunisia
A 12	Bridge at Maktar, Maktar	Oued Djilf	Tunisia
A 13	Bridge at Souk el Arba, Souk el Arba	Oued el Henja	Tunisia
A 14	Bridge at Ippona,	Oued Bou Djim	Tunisia
A 15	Bridge at Theveste, Theveste	Oued Meskiana	Tunisia
A 16	Bridge at Ain Hedja, Ain Hedja		Tunisia
A 17	Bridge at Tebourso, Tebourso	Oued el Ahmar	Tunisia
A 18	Bridge at Ain Younes, Ain Younes		Tunisia
A 19	Bridge at Sousse, Hammamet	Oued Bou Zid	Tunisia
A 20	Siliana Bridge, Maktar	Siliana	Tunisia
A 21	Bridge at Adpertusa, Adpertusa	Oued Chafrou	Tunisia
A 22	Bridge at Civitas Avensis, Souk el Khemis		Tunisia
A 23	Bridge at Kairouan, Kairouan	Oued Nebana	Tunisia
A 24	Bridge on Kerkenna, Kerkenna Island		Tunisia
A 25	Bridge at Medjez, Medjez el Bab	Oued Zargua	Tunisia
A 26	Bridge at Uthina, Oudna		Tunisia
A 27	Bridge at Sufetula, Kasserina	Oued Sbeitla	Tunisia
A 28	Bridge at Tolemaide, Tolemaide,	Uadi Ziuana	Libya

Near East

	Name, location	River	Country
NE 01	Bridge at Maameltein, Beirut		Lebanon
NE 02	Bridge at Tyrus, Tyre	Nahr Abou-el-Assouad	Lebanon
NE 03	Jisr-el-Majami, Samaria	Jordan	Palestine
NE 03	Jisr-el-Siq		
NE 05	Jisr Jindas, Lydda		Israel
NE 06	Five bridges, Lydda, Haifa, Tel Aviv		Israel
NE 07	Bridge at Er-Rukkad, Quneitra	Wadi Zedi	Syria
NE 08	Bridge at Er-Rukkad, Quneitra	Wadi Qudana	Syria
NE 09	Three bridges, Bosra	Wadi Zedi	Syria
NE 10	Nimre Bridge, Bosra		Syria
NE 11	Bridge at Aleppo, Aleppo	Afrin	Syria
NE 12	Two bridges, Aleppo	Saboun-Souyou	Syria
NE 13	Bridge at Assuan, Assuan	Nile	Egypt
NE 14	Bridge at Mossul, Eski Mossul	Tigris	Iraq
NE 15	Bridge at Shushtar, Shushtar	Karum	Iran (IM)

7. Case studies

Between 1987 and 2000 the author visited many Roman bridges to record their current condition by means of photographs and by taking measurements. He studied the relevant sources and combined their findings in his description of each bridge.

To facilitate the quick evaluation of each bridge and comparisons among several of them a standardized introduction gives the basic technical data such as

river crossed,
road served,
location,
province,
length,
width,
height,
number and sizes of spans,
pier width,
materials used,
built by,
time of construction,
date visited,
H/R value.

The significance of the H/R value is explained in chapter 5.5.4, page 77.

1. View of the rebuilt bridge.
2. Downstream view of pier.
3. Upstream view of the original pier with its break-water and the concrete arches with the preserved stones attached.

IN 16. Ponte sul Rubicon

River: Rubicon, road: Via Aemilia, location: Savignano, province: Emilia-Romagna. Length: 26.2 m, width: 7.9 m, height: 8 m, spans: 3 x 6.5 m, material: limestone, ratio of voussoirs to radius: 1.10/3.25 m = 0.3385, pier width: 2.4 m. Built under Augustus (?), time of construction: 1st c. BC. Visited: May 1992.

A 4.1 m wide pavement plus two 1.2 m sidewalks and two 70 cm parapets add up to the 7.9 m width of the bridge. The arches start 1.8 m above the water level and consist each of 15 voussoirs of about 63 x 100 cm. The core was *opus caementitium*, suggesting the 1st century BC as construction time. In 1944 the bridge was dynamited. It was rebuilt in reinforced concrete, which was faced with the original stones to obtain the previous appearance.

The Rubicon is famous for Caesar's decision to move on to Rome with his army (alea iacta est). Savignano is 10 km east of Cesena.

Ref.: Baldoni 1979, p. 61; Balance 1951, p. 115; Gazzola 1969, p. 54.

IN 19. Ponte d'Augusto

1. General view of the bridge.
2. Up-river view showing worn cutwaters.
3. Aedicula between arches.

River: Marrecchia, road: Via Aemilia, location: Rimini, province: Emilia-Romagna. Length: 62.6 m, width: 8.65 m, height: 8 m, spans: 8.9 + 8.9 + 10.6 + 8.7 + 8 m, material: travertine, ratio of voussoirs to radius: 1.25/5.30 m = 0.2358, pier width: 3.2. Built under Augustus/Tiberius, time of construction: 1st c. AD. Visited: April 1989.

Since 187 BC the Via Aemilia continued the Via Flaminia from Rimini north-westwards. Augustus began a bridge over the *Ariminus*, which was completed under Tiberius in AD 20, hence the bridge is today known as Ponte Tiberio. It is one of the most classical Roman bridges, heavy set and well proportioned, with exactly semicircular arches resting on pile supported piers. Today the river is reduced to a pond, connecting with a small harbour on the sea. In 552 Usdrilla, king of the Goths, destroyed part of the bridge, to stem the Byzantine advance, it was soon after repaired. In the 14th century flooding damaged the bridge, community and church had it restored, as did Pope Innocent X in 1680.

The 4.9 m wide pavement (square stone slabs on gravel and pozzuolana earth) handles the modern traffic with ease, leaving enough space for sidewalks and parapets. The carefully cut voussoirs are 1.25 m high by up to 1.8 m deep and 50 cm wide, weighing well above 2 tons each. The bridge carries a cornice with a dental pattern projecting 60 cm, the spandrels above the piers are embellished with aedicule 1.6 x 2.4 m, which once contained statuary. This gives the bridge an architectural quality.

Ref.: Ashby and Fell 1921, pp. 190 and 215; Gazzola 1969, p. 73; Merckel 1902, p. 291.

1, 2. General views of the bridge.

IN 31. Ponte sulla Dora, Ponte Vecchio

River: Dora, road: Via Aemilia, location: Ivrea, province: Piemont. Spans: three, material: limestone, brick. Visited: April 1989.

This bridge is at the entrance to the Aosta valley, which was the preferred passage to Gallia. A military road should have been built in the 2nd c. AD to prepare for Caesar's conquest of Gaul. The bridge, which is also called Ponte Vecchio or Ponte Romano is built in brick with concrete backfill (*opus latericium*), the piers in limestone. It is subject to severe flooding by the water descending from the Alps and needs frequent repairs.

Ref.: Blake 1947, p..218; Gazzola 1969, p. 96.

IN 35. Ponte sul Lys at St. Martin

River: Lys, road: local, location: St. Martin, province: Aosta. Length: 62 m, width: 5.82 m, height: 25 m, span: 36.65 m, material: local stone, ratio of voussoirs to radius: 0.70/18.32 cm = 0382. Built under Augustus (?), time of construction: 1st c. BC. Visited: April 1986.

This bridge is an outstanding example of Roman construction skill. It has the biggest known span, achieved with one segmental arch of slightly parabolic shape. The rise of the arch is 12 m, its center is about 6 m below the impost level. In general, parabolic arches with a slight hump appear during late imperial times. Parabolic means that every voussoir has a slightly different angle, which explains the variable width (20 to 80 cm) of the stones. The bridge rests on bedrock, the northern impost being lower, meaning 42 voussoirs to the top, against 38 voussoirs on the southern half. They were laid in five 80 cm wide arches with concrete fill between and above. Most likely they were built step by step, reusing the formwork. The abutment walls are built with polygonal masonry, leveled off every meter by a projecting course of ashlars, adding a light touch to the structure. Some voussoirs are have received metal anchors to the adjacent masonry. The 60 cm parapet rests on a projecting molding, leaving 4.62 m for the pavement of stone slabs or pebbels. It seems, the bridge had to be rebuilt under Augustus, replacing an earlier bridge from the 2nd c. BC.

Ref.: Blake 1947, p. 217; Gazzola 1969, p. 97.

1. General view of the bridge.
2. Southern abutment of the bridge.
3. Northern abutment of the bridge.
4. Pavement at the crown of the bridge, with Baroque gate in the background.

page 101
1. View of the bridge by Aubert (1860).
2. Longitudinal section of the bridge, established 1996 by M. Döring.
3. Today's crossing at aqueduct level.
4. Cross section of the bridge by Döring. Left: the antique bridge, right: present-day condition with passage at the level of the aqueduct.
5. General view.

IN 36. Ponte del Pondel

River: Grand Evia, road: passage and aqueduct, location: Val di Cogne, province: Aosta. Length: 60.46 m, width: 2.26 m, height: 66 m, spans: 14.24 m, material: local gneiss, ratio of voussoirs to radius: 1/9.65 m = 0.1036. Built under C. Avilius and C. F. Caimus, time of construction: 3 BC. Visited: July 1991.

One of the most daring Roman bridge structures, the principal reason for its great height is its function as aqueduct, with the pedestrian passage as collateral benefit. The aqueduct supplied water to a nearby mining operation and was financed by the owners of the mine. It further down also irrigates an agricultural area. The arch consists of 53 voussoirs, 1.2 m high by 80 cm wide, the keystone being 2 m high. The abutments are 46 m above the river, the arch reaches a level of about 54 m, too low for the aqueduct. To gain more height at minimal load, two walls with windows were built to support the aqueduct, creating a 1 m wide by 3.88 m high passage covered with slabs of gneiss. Above, the water conduit was 85 cm wide and for inspection purposes 1.95 m high, again covered with gneiss, creating a topmost passage of 1 m.

Also the intermediate space is accessible and could have served as walkway or for storage.

Ref.: Döring 1998, p. 127; Gazzola 1969, p..98; Heinz 1989.

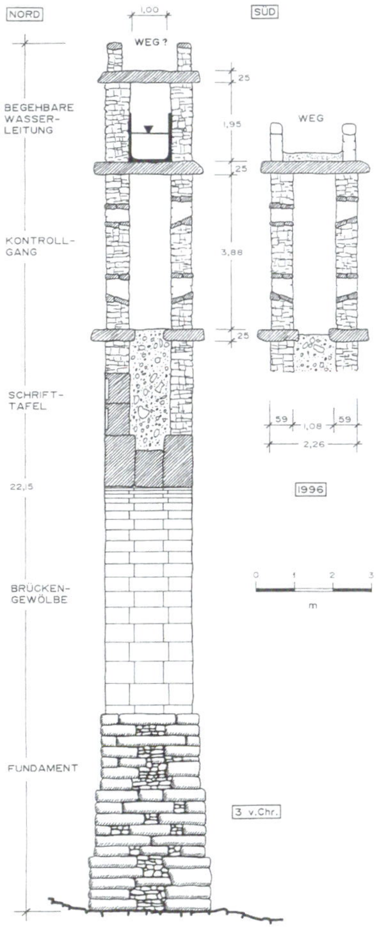

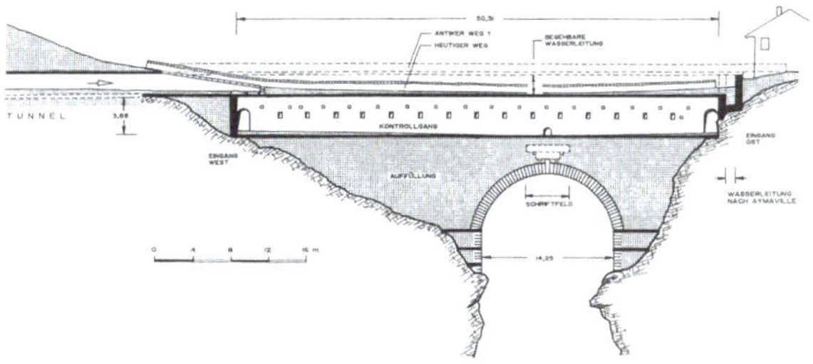

IN 37. Ponte sul Torrente Buthier

River: Bauthegius, road: Bernardino, location: east of Aosta, province: Aosta. Width: 5.9 m, span: 15.05 m, material: gneiss, ratio of voussoirs to radius: 95/752.5 cm = 0.1263. Built under Augustus (?), time of construction: 1st c. BC. Visited: July 1991.

The river under the bridge has long ago been diverted, the structure is now a Roman monument in the town. The three abutment stones are 80 x 80 cm and up to 2 m long. Thus the arch starts at 2.4 m. It has 33 vousoirs 45 x 96 cm. The arch is distinguished by a 1.5 m wide keystone, three times that of a voussoir. The road leads from Ivrea to the small Saint Bernard Pass and passed through a triumphal arch built by Augustus on occasion of the founding of the Roman colony Augusta Praetoria Salassorum in 25 BC.

Ref.: Gazzola 1969, p. 99.

1. The arch in the town.
2. Detail of the abutment.
3. The arch of Augustus, protected by a modern roof.

1. Second bridge of Val Quazzolano.

IN 41. Secundo Ponte della Val Quazzola

River: Torrente Quiliano, road: Julia Augusta, location: Quiliano (Savona), province: Liguria. Length: 7.6 m, width: 5.57 m, height: 8 m, span: 7.6 m, material: limestone, concrete, ratio of voussoirs to radius: 90/380 cm = 0.2368. Time of construction: 2nd c. AD. Visited: July 1991.

The long coastal road from Italy to France is known as Via Julia Augusta, it was begun by the Republic. An older branch of Via Julia Augusta went from Turin westward over the Mont Genèvre Pass directly to Gallia Narbonensis. The coastal road was finished under Hadrian in AD 124 and today is generally referred to as Via Aurelia. It usually skirted the mountains, but had to cross the many streams. For these bridges a standard design of relatively small span was adopted, employing flat stones of 10 to 12 cm thickness and 40 to 50 cm height. To reach the necessary strength two arches were put on top of each other. The cutting of wedge shapes was avoided. Instead the thickness of the mortar joints increased from the bottom up to give the radiating position of the voussoirs. There is a number of such bridges in the Val Quazzola.

This bridge is an example of this technique. Most likely being built by army engineers it was a fast method, avoiding the quarrying and transport of big loads.

Ref.: Gazzola 1969, p. 106.

IN 43, Quinto Ponte di Val Ponci – Ponte delle Fate

River: Val Ponci, road: Julia Augusta, location: Finale, province: Liguria. Width: 6.56 m, height: 6 m, span: 6.55 m, material: limestone. Built under Augustus / Hadrian, time of construction: 1st / 2nd c. AD. Visited: July 1991.

The fifth bridge in Val Ponci is in good condition and used for local traffic. The arch consists of two ranges of relatively thin stone slabs, each 12 to 15 cm thick and 45 cm high, 100 slabs per ring. The curvature of the rings can be achieved by variable thickness of the mortar joints, which is highly economical. Six ranges of heavy blocks, 50 cm high by up to 250 cm long are used for the abutments, which have long guide walls (27.8 m to the south) to direct the water and protect against undercurrents. Spandrel walls are made of small 12 to 15 by 15 to 45 cm limestone backed by concrete. The massive substructure is needed against the raging torrent coming down from the Alps. All in all this bridge is an excellent example of various building techniques combined to full advantage. The bridge can be reached by car.

Ref.: Gazzola 1969, p. 110.

1. View from southwest, exhibiting some decay in the masonry and a portion of a guide wall.

1. Exposed Roman concrete of the 3 m roadway.
2. Typical arch.
3. Southwest view of bridge along the modern highway.

IN 48. Ponte Lungo

River: Bauthegius, road: Via Julia Augusta, location: Albenga, province: Liguria. Length: 148 m, width: 3.5 m, spans: ten x 6.6 to 8.4 m, material: limestone, ratio of voussoirs to radius: 80/420 cm = 0.1905. Built under Augustus (?), time of construction: 1st c. BC. Visited: July 1991.

In comparison to neighboring bridges, which have a width of 5.5 to 6.2 m the Ponte Lungo is quite narrow. Some question therefore if it is a Roman bridge at all. It has two superposed arches with voussoirs 60 to 80 cm high at the bottom and 30 cm above, which speaks for Roman work, while the narrow net paving of 3 m speaks against. It is crossing the wide river bed close to the Mediterranean coast. Now silted up it hardly rises more than 3 m above the terrain, a modern road running parallel to it. The pavement is lost and we can see the *opus caementitium* composed of stones laid vertically in mortar.

Ref.: Gazzola 1969, p. 112.

IM 1. Milvius Pons

River: Tiber, road: Via Flaminia, location: Rome. Length: 150 m, width: 7.5 m, height: 15 m, spans: 9 + 18 + 18 + 9 m, material: tuff, travertino, gabino, ratio of voussoirs to radius: 1.2/9 m = 0.1333, pier width: 7.5 m. Built under Flaminius (?), time of construction: 3rd c. BC. Visited: July 1992.

The Via Flaminia, built in 220 BC by Censor Gaius Flaminius was the most important northward road. The road was described by Ashby and Fell (1921, its bridges by Balance (1951) and Blake (1947). It has the greatest number of bridges of Roman roads.

Milvius Pons, also called Ponte Molle, is situated in northern Rome. It is known through the victory of Constantine I over Maxentius on 28 December 312 AD, an event connected with the adoption of the cross by Christian iconography (*in hoc signo vinces*). This part of the Via Flaminia was begun in 220 BC, and the bridge is first mentioned by Livy in 207 BC. In 174 BC it is recorded as a timber bridge on stone piers, in 142 the timber was replaced by stone arches. Around 109 BC the Censor Aemilius Scaurus ordered repairs or even reconstruction. In 27 BC Augustus commissioned maintenance works for the Via Flaminia, excluding Milvius Pons and Minucius Pons, which must have been in good condition. To commemorate the works a triumphal arch was built in the middle of the bridge.

Today's condition stems from a restoration by Pope Pius VII in 1805. In 1849, when hi defended Rome, Garibaldi had one arch destroyed. It was rebuilt within a year. Today the bridge serves for pedestrians only, but in World War II heavy tanks rolled over it.

Nowadays the northern abutment is hidden, also the seventh arch of the north ramp. The square 7.50 m piers, cutwaters and floodways are original and consist of a tuff core faced with travertine. Arches 1 to 6 seen from north were variably restored, arch 1 having a brick vaulting, 2 and 3 gabino vaults with facing travertine voussoirs 60 x 120 cm. The fourth arch consists of two layers of travertine, topped by brickwork. Arches five and six are entirely built with brick. Gabino is a volcanic rock quarried at the ancient town of Gabii.

Ref.: Ashby 1927, p. 247; Gazzola 1963, pp. 33–35; Merckel 1902, p. 289.

1. View from south. Brick was used for most of the repairs.

1. View from the west.

IM 2. Fabricius Pons

River: Tiber, road: urban, location: Rome. Length: 61.25 m, width: 6 m, height: 15 m, spans: 25 + 24.7 m, material: tuff, peperino, travertino, gabino, ratio of voussoirs to radius: 1.5/12.4 m = 0.121, pier width 9.5 m. Built under L. Fabricius, time of construction: 62 BC. Visited: July 1992.

Fabricius Pons, commonly called Quattro Capi, is the only completely preserved Roman bridge in Rome. It is one of the most beautiful Roman bridges. Built at the same time as the Cestius bridge (IM 5), it connects the Tiber isle with the left bank of the river. Inscriptions identify the builder as *curator viarium* L. Fabricius and the date of construction. A second inscription records repairs in 21 BC by Q. Lepidus and Consul M. Lollius.

The bridge is easy to measure up since the balustrade has metal markers indicating the cen ter of arches or piers. The pier between the two arches has a big floodway. The arches are faced with white travertine, each with 61 voussoirs of 65 x 150 cm. Vaults, piers and abutments are of gabino. Originally the bridge had a bronze railing and was embellished with statues of Hermes, giving the bridge the popular name Ponte dei Quattro Capi. Pope Innocent XI (1676–89) replaced the railing with a stone parapet and used the metal for coinage.

Ref.: Blake 1947; Chevalier 1976; Gazzola 1963, pp. 41 f.; Heinrich 1983, p. 39; Lugli 1957, p. 197 (details about stone masonry); Merckel 1902, pp. 281 f.; Nash 1968; Platner and Ashby 1929.

IM 3. Aemilius Pons, today Ponte Rotto

1. The inaccessible remains of the bridge.
2. 17th-century view by van Nieulandt (1584–1635).

River: Tiber, road: urban, location: Rome, province: Rome. Height: c. 15 m, spans: 6 x up to 24 m, material: tuff, travertino, gabino, ratio of voussoir to radius: 1.2/9 m = 0.1333. Built under Lepidus and Nobilior, time of construction: 179 BC. Visited: July 1992.

It is said to be the first bridge with stone piers in Rome, appropriately named Pons Lepidus. Being commissioned by Censor M. Aemilius Lapidus the name served both origins (Jurecka 1986). In 142 BC Scipio Africanus and L. Mummius replaced the wooden spans with peperino stone arches. Augustus restored it with travertine.

In 1557 a flood destroyed the bridge, Pope Gregory XIII rebuilt it, but in 1598 the eastern part collapsed again. Later an iron suspension bridge was superposed, which was removed in 1885, together with the western arches. Today only two piers and one arch of the southern part remain, explaining today's name »broken bridge«.

Ref.: Blake 1947; Blake and Bishop 1973; Chevallier 1976; Gazzola 1963, p. 33; Jurecke 1986, p. 67; Nash 1968; Platner and Ashby 1929.

1. Today's bridge with Castel Sant'Angelo (Hadrian's Mausoleum).
2. View of the original bridge from *urbis aeternae vestigia* by Piranesi, 1786.

IM 4. Aelius Pons (Angels' Bridge)

River: Tiber, road: urban, location: Rome, province: Rome. Length: 135 m, width: 10.95 m, height: 15 m, spans: 2 x 7.5 + 3 x 18 + 2 x 7.5 + 3.5 m, material: peperino/gabino concrete, travertino. Built under Hadrian, time of construction: 136 BC. Visited: July 1992.

When Emperor Hadrian built a mausoleum for himself and his family on the right bank of the Tiber, he added an access bridge with three arches of 18 m spans, called Pons Hadriani. Changes of the river necessitated the extension of the bridge to eight spans. It had a 33 m ramp on the city side and a 22 m ramp on the mausoleum side. In 1527 Pope Clement VII placed statues of Peter and Paul at the east end (paid from the bridge toll), later augmented by figures of the evangelists. Since the statues had decayed, Pope Alexander VII commissioned new Baroque statues of angels by Bernini, hence the name of the bridge.

In the course of a Tiber regulation at the end of the 19th c. the old bridge was replaced by a new one with five arches, each with a span of 18 m, which increased the flow capacity. The construction works revealed many details of the bridge's history.

Ref.: Blake and Bishop 1973: Gazzola 1963, p. 131; Merckel 1902, p. 285; Neuburger 1929, p. 476; Platner and Ashby 1929.

IM 5. Cestius Pons

River: Tiber, road: urban, location: Rome, province: Rome. Length: 68.8 m, width: 8.9 m, height: 15 m, spans: 20.2 + 25.5 +21.4 m, material: tuff, travertino, gabino, ratio of voussoir to radius: 1/12.75 m = 0.0784. Built by Lucius Cestius, time of construction: 46 BC. Visited: July 1992.

The bridge was built by the *curator viarum* Lucius Cestius in 46 BC while Caesar battled in Spain. Repaired in AD 152 by Antonius Pius and rebuilt after a flood in AD 370 by Valentian I and Gratian it was named Pons Gratiani. It had a 25 m middle arch flanked by 5.8 m arches. During the 19th century building rage it was torn down and newly built. The middle arch was retained its original size, 347 of the previous 563 travertine blocks of the Pons Gratiani were reused. Marble from the theatre of Marcellus built under Trajan was also used. The middle arch has 59 voussoirs 68 x 100 cm, the side arches have 55 voussoirs 60 x 100 cm. The demolition revealed an intricate system of iron clamps set in lead to reinforce the masonry.

Merckel considered this to invite the penetration of water, leading to frost damage and rusting – which would be the case if the leading was not properly done.

Ref.: Blake and Bishop 1973; Gazzola 1963, p. 42; Heinz 1988, p. 62; Merckel 1902, p. 279; Nash 1968; Platner and Ashby 1929; RE XXI, 2470.

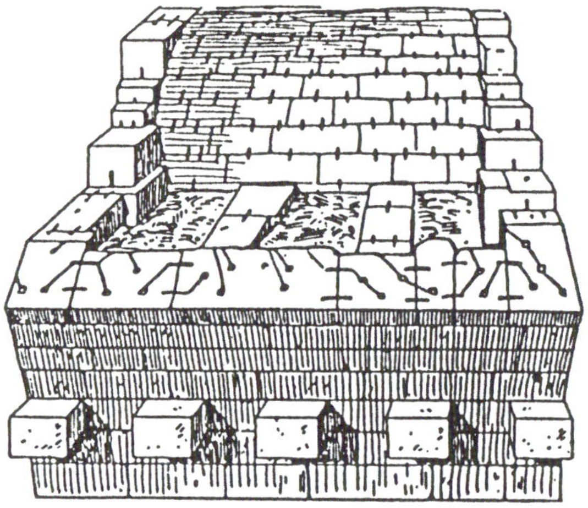

1, 2. System of metal ties inserted by the Romans.
3. View from south.

1. View of part of the causeway.
2. Detail of masonry.

IM 10. Viadotto di valle Ariccia

Road: Via Appia, location: 1 km south of Ariccia, province: Latium. Length: c. 300 m, width: 8.2 m, height: 11.2 m, spans: 2 x 3 m, material: peperino, concrete, tuff, time of construction: 2nd c. BC. Visited: July 1992.

This causeway consists of 15 ranges of 57 cm high blocks up to 1.65 m long. It includes two vaulted culverts for drainage. It rises along the flanks of a crater and is an example of the Roman's mastery over nature's obstacles, which was already admired by Piranesi, Canina and Schinkel for the quality of its execution. Juvenal (Sat. IV) relates, how beggars took advantage of the slow uphill traffic to collect alms. Today the structure is difficult to find, it begins 1 km south of Ariccia, most of it is hidden under the growth of a private vineyard. 100 m northward is a small traffic circle with a limestone arch from the Via Appia, called *basto del diabolo*, meaning the devil's saddle.

Ref.: Adam 1984, p. 307; Blake 1947, p. 212; Esch 1997; Gazzola 1963, p. 17; Lugli 1957, p. 47.

IM 13. Ponte di Tre Ponte

1. Heavy masonry for arch and parapet
2. Via Appia and bridge.
3. Part of the inscription.

River: Ninfa, road: Via Appia, location: Treponte, province: Latium. Length: 31.6 m, width: 12.5 m height: 5 m, spans: 2.25 + 2 x 5.6 + 2.25 m, Material: large tuff blocks, ratio of voussoir to radius: 1/2.8 m = 0.3571, 60/113 cm = 0.5310. Built under Nerva/Trajan, time of construction: 2nd c. AD. Visited: July 1992.

In 312 BC Appius Claudius built the Via Appia, which passed between Cisterna di Latina and Terracina through 40 km of the Pontine Marshes. Parts had to be crossed by boat. Trajan drained the swamps and built bridges to let the water pass. This bridge, most likely founded on piles, is in good condition and still carries today's traffic. In its middle is a 1 x 2 m tablet indicating Trajan as builder and subsequent renovations (CIL X, 6819). Another inscription mentions repair works under Constantine. The width of the bridge consists of 1 m cornices, 2 x 75 cm parapets, 2 x 1.3 m sidewalks and 7.4 m pavement.

Ref.: Gazzola 1963, p. 23; Quilici 1989, p. 8.

1. Reconstructed view of the bridge by Colin O'Connor.
2. Different springing levels of arches to accommodate increasing height. Note the concrete core.
3. Remaining 20 m span.

IM 27. Ponte di Augusto

River: Nera, road: Via Flaminia, location: Narni, province: Umbria. Length: 160 m, width: 8 m, height: max. 30 m, spans: 16 + 18 + 32 + 20 m, material: limestone concrete, travertino. Built under Augustus, time of construction: 27 BC. Visited: April 1989.

In 550 Prokop of Kaisareia describes the bridge in de aedificiis, saying he never saw a taller vault. In the Middle Ages Hermann von Reichenau writes about this monument to the art of building. The inclined bridge has variable spans to accommodate the increasing height. The piers, which do not have cutwaters, most likely were weakened through erosion. Already Augustus had to rebuild the structure: *potemque mirae magnitudinis in sublimitate fundavit, et intra triginta annos explevit; et signavit carmine, Caesar Augustus de Manubiis Narn"iensis Patronus* – »He initiated a majestic bridge of astonishing size and completed it within 30 years. In an inscription he indicated: From the proceeds of military expeditions by Caesar Augustus, protector of Narni.« Parts of the bridge collapsed again in 800 and 1053 and were rebuilt.

Apparently the structure was too daring to survive complete, indeed it was unique among Roman bridges. One 20 m arch remains at the south end, only the lower parts of the piers survived, one of them toppled over by erosion. The abutment wall on the north side, which was pierced for a railroad line, shows at the springing of the arch projecting ribs, most likely for visual effect.

Ref.: Ashby and Fell 1921, p. 169; Balance 1951, p. 91; Blake 1946, p. 213; Esch 1995, p. 85; Gazzola 1963, p. 57; Merckel 1902, p. 291; Niessen 1902.

IM 32. Ponte Fonnay

1. The culvert passes under the road at an angle of 60°.
2. Fifteen voussoirs form the 3.5 m wide arch.

River: brook, road: Via Flaminia, location: Massa Martana, province: Umbria. Length: 13.5 m, width: 16 m, height: 8 m, spans: 3.5 m, material: tuff, ratio of voussoir to radius: 60/170 cm = 0.3529. Built under Augustus, time of construction: 1st c. BC. Visited: April 1989.

Excellent workmanship distinguishes this bridge. To reach the road level 8 m above, the narrow arch has been stilted by 5 m. The investment of so much masonry to achieve a uniform slope of the road will have impressed the local population. The culvert, which crosses the road at an angle of 60° resulted in a 16 m long vault. 10 m wing walls of 40 to 60 cm high and 80 to 100 cm long stones direct the water to the culverts. Several additional culverts and walls have been uncovered.

The structure is difficult to reach, at 3 km north of Acquasparta and 2 km south of Massa Martana one turns west, a track leads about 500 m further to the bridge.

Ref.: Gazzola 1969, p. 59.

1. General view of church.
2. Remaining arches below the nave of the church.

IM 33. Ponte di Giovanni de Butris

Road: Via Flaminia, location: Massa Martana, province: Umbria. Width: 8.7 m, spans: 2 x 3.45 m, material: limestone, ratio of voussoir to radius: 60/172 cm = 0.3478, time of construction: Republic. Visited: April 1989.

This is an example of the appropriation of a bridge for other uses: A Romanesque church has been built upon the Roman bridge. Two arches cross a brook at 15°, each has 9 voussoirs 60 x 75 cm. Again the excessively massive design was to impress people. As historical irony the work of the pagan Pontifex Maximus supports the building of the papal Pontifex Maximus. Today the brook flows at a distance of 100 m and the bridge served as foundation of the church and material for its construction.

Ref.: Esch 1997; Gazzola 1969, p..60.

IIM 34. Ponte del Diavolo

River: drain, road: Via Flaminia, location: Bastardo, 12 km sw of Bevagna, province: Umbria. Length: 9.1 m, width: 14.7 m, height: 3.5 m, spans: 3.3 m, material: limestone, time of construction: Republic. Visited: April 1989.

The small structure surprises by its massive execution and great width. 60 by 100 cm voussoirs for a 3.3 m span are excessive, as is the width of the bridge – hence the attribution to the devil. The arch is 100 m off the modern Via Flaminia.

Ref.: Gazzola 1969, p. 61.

IM 38. Quarto Ponte in Foligno

Location: Foligno, via delle Grazie, province: Umbria. Length: 50 m, width: 4.5 m, height: 10 m, spans: 9 + 12 + 9 m, material: limestone, travertino, ratio of voussoir to radius: 50/450 cm = 0.1111. Built under Augustus, time of construction: 1st c. BC. Visited: April 1989.

Medieval city walls have been built over this bridge, the previous river has shifted northward. Today's bridge consists of two arches. All four bridges in Foligno are similar.

Ref.: Gazzola 1969, p. 63.

1. Ponte del Diavolo. The structure belongs to a farm.
2. Quarto Ponte in Foligno.

1. View showing angular abutment walls.
2. General view from west.
3. View from east.

IM 49. Ponte Grosso Burano

River: Burano, road: Via Flaminia, location: 3 km north of Cantanio, province: Pesaro. Length: 36 m, width: 6.5 m, height: 9 m, spans: 2 x 7 m, pier width: 5.6 m, material: limestone, pietro corgnola, ratio of voussoir to radius: 75/350 cm = 0.2143. Built under Augustus, time of construction: 1st c. BC. Visited: April 1989.

The bridge is between Foligno and Pésaro, 7 km south of Cagli and 3 km north of Cantiano. It crosses the rough ravine of the Burano. The northern arch is fully original, the southern arch a bit repaired. A massive pier with pointed projections against the stream joins two arches. The abutment walls begin at right angle to the bridge, then open up at 15° to direct floodwater to the stream.

The Burano is known for flash floods. Masonry work is excellent, consisting mainly of flat, 20 cm thick, quarry face stone, which occasions the name of the bridge. Originally being of 10 m Flaminia road width, it seems three or four arches on the Westside collapsed and narrowed the bridge by 3.5 m. The parapets were shifted inward in medieval times, reducing the width of the pavement.

Ref.: Balance 1951, p. 107; Gazzola 1969, p. 83.

IM 50. Ponte Manlio

River: Bosso, road: Via Flaminia, location: Cagli, province: Pesaro. Length: 200 m, width: 8.85 m height: 8 m, spans: 11.7 + 4 m, pier width: 5.6 m, material: limestone conglomerate, grigna, corgnola, ratio of voussoir to radius: 1/5.83 m = 0.1715,time of construction: Republic. Visited: April 1989.

This structure is very well preserved. The massive blocks of conglomerate are very old, the locals claim, the bridge was built by the Etruscans. Most likely it was built in the 3rd c. BC under Roman auspices. The name derives from an inscription (CIL XI, 753), which indicates also repairs under a certain M. Allius Tyrannus. Three construction periods can be detected: Early on, the main arch and parapet, then an extension under Augustus using pietra corgnola like in IM 49. Lastly renovations under Trajan and Hadrian (2nd c. AD).

The principal arch is largely silted up, it has a keystone 1.25 m high, 1 m deep and 1 m wide, weighing about 4 tons. The spandrel blocks are up to 2.5 m long. The parapet blocks project 40 cm. There are again tilted approach walls in low-range masonry, only the northern part is preserved. The bridge is at the north exit from Cagli.

Ref.: Balance 1951, p..109; Gazzola 1969, p..46; Radke 1971, p..22.

1. Detail of main arch. The voussoirs have not moved at all in 2300 years!
2. View from west.
3. Note the generous width of the bridge.

1. View of the causeway, being a classified monument it is protected by a metal roof!
2. The northern 2.8 m arch with its massive voussoirs.

IM 56. Ponticello dell Abbazia

River: Burano, road: Via Flaminia, location: 3 km north of Cantanio, province: Pesaro. Length: 36 m, width: 6.5 m, height: 9 m, spans: 2 x 7 m, pier width: 5.6 m, material: pietro corgnola, ratio of voussoir to radius: 75/350 cm = 0.2143. Built under Augustus?

The bridge is between Acqualagna and Fossobrome, near the Abbazia San Vicento at the southern entry to the Gola (gorge) de Furlo. It is part of a 50 m long causeway along the Candigliano river, built to maintain a straight road alignment. A little shift closer to the hillside would have meant a small curve, but rendered the expensive construction unnecessary. So much for Roman forwardness.

The two arches 30 m apart are greatly overdimensioned, 1 m high voussoirs for less than 3 m span are excessive, as is the width of the road. The *corgnola* stone is a very cohesive conglomerate of limestone chippings, contributing to the excellent condition of the masonry.

Close by the construction of the road bed can be seen. It has an 1 m thick anti-freeze layer of limestone chips of 5 to 25 mm grain size. The road continues up to the Furlo pass, featuring the 38 m long Tunnel of Vespasian, built AD 77.

Ref.: Gazzola 1969, p. 81.

IM 72. Piccolo Ponte di Nona

1. Arch against a pier of the new bridge.

Road: Via Praenestina, location: 15 km east of Rome, province: Latium. Span: 4 m, material: peperino, time of construction: 173 BC. Visited: July 1992.

The bridge is preceding the great bridge of Nona, which passes 13 m above it, avoiding a descent to the river. Maybe religious considerations kept the old bridge in place. It was mentioned by Livy as built in 173 BC. The bridge is 20 miles east of Rome along the Via Praenestina.

Ref.: Blake 1947, p 211; Gazzola 1963; Radke 1971, p. 66; RE XXI, 2432.

1. Overall view from north.
2. Two arches meeting upon an impost.
3. Close-up view of arches.

IM 73. Grande Ponte di Nona

River: Marrana, road: Via Praenestina, location: Nona, province: Latium. Length: 110.5 m, width: 11.53 m, height: 17 m, spans: 7 x 6.45 m, pier width: 5.6 m, material: peperino, gabino, ratio of voussoir to radius: 1.5/3.225 m = 0.4651. Built under Sulla (?), time of construction: 1st c. BC. Visited: July 1992.

Arguably the most impressive Roman bridge in the vicinity of Rome, it consists of seven equal arches and large, uniformly 50 cm high rusticated gabino and peperino masonry with a concrete core. The height of the piers varies from 6 to 17 m with the configuration of the valley. The 1.5 m high voussoirs have accentuated keystones made of travertine. The bridge still serves modern traffic without restriction.

It should be noted, that the measurements by the author differ from those of others. He gives the length as 110.5 m, against 72 m by Gazzola, 71 m by Ashby and 87.3 m by Merkel. This may be explained by changing abutment conditions.

Ref.: Ashby 1927; Gazzola 1963, p. 44; Merckel 1902, p. 291; Lugli 1957, p. 198.

IM 74. Ponte Amato

1. View in cleaned condition.
2. Condition at present.

River: Fosso dell'Infernatio, road: Via Praenestina, location: Nona, province: Latium. Length: 310 m, width: 5.65 m, height: 10 m, spans: 6.5 m, material: gabino, ratio of voussoir to radius: 1/3.25 m = 0.3077. Built under Sulla, time of construction: 1st c. BC. Visited: July 1992.

Originally an aqueduct of the Acqua Claudia, it was enlarged to a road bridge. Similar to the Nona bridge it has massive masonry of up to 65 cm high gabino blocks. The arch is part of a long causeway across the Fosso dell'Infernatio, which has been built in *opus quadratum*, 17 ranges of ashlars 56 to 65 cm high. The name derives from Conte di Segni Amatus, who had it restored. Today it is hidden under growth on private property.

Ref.: Adam 1984, p. 309; Ashby 1927; Blake 1947, p. 212; Gazzola 1963, p. 26; Radke 1971, p. 67; RE XXI, 2433.

1. General view of the bridge.

IM 89. Ponte Lucano

River: Anius, road: Via Tiburtina, location: 4 km west of Tivoli, province: Latium. Length: 61.7 m, width: 7 m, height: 4.35 m, spans: 5 + 8.4 + 9.2 + 6.6 m, piers 6 to 7 m, material: tuff, travertino, ratio of voussoir to radius: 65/650 cm = 0.1000. Built by M. Plautius, time of construction: 2nd c. BC. Visited: July 1992.

Of the five arches of the original bridge only four remain. According to Prokopius (AD 500–560) Totila destroyed all bridges between Rome and Tivoli. They were rebuilt by the Byzantine general Narses (490–575). This bridge was built in tuff and faced with travertine. Due to pondage the level of the Anius river has risen covering the lower part of the bridge, which makes it difficult to establish exact measurements. Two 30 cm thick parapets reduced the paving to 6.38 m. The bridge has been restored with bricks and plastered, changing its character entirely. Nearby is the 2nd-century tomb of Consul Plautinus Silvanus who most likely had the bridge built.

Ref.: Ashby 1927; Blake 1947; Gazzola 1963, p. 49; Merckel 1902, p. 289; Paget 1973; Radke 1971, p. 66.

IM 91. Ponte Sambuco

River: brook, road: Via Salaria, location: 4 km south of Rieti, province: Latium. Length: 64 m, width: 6.4 m, height: 8 m, span: 3.8 m, material: limestone, ratio of voussoir to radius: 60/190 cm = 0.3158. Built under Augustus, time of construction: 1st c. BC. Visited: April 1989.

This bridge is a small opening in a long and tall ramp to permit a straight road, again a rather superfluous effort to achieve an impressive expression of Roman power. The short arch has 15 voussoirs of 40 x 70 cm, the lengths of the ramp blocks vary between 90, 120, 180 and 210 cm. The road has been abandoned, from time to time the overgrowth is cleared. Massive masonry keeps the structure in good shape.

To find the bridge take the N4 from Rieti 4 km southwest, then a small road for 1 km eastward and search in the fields.

Ref.: Esch 1997; Gazzola 1963, p. 82; Lugli 1957, p. 356; Quilici 1994, pp. 118 f.

1. The rough masonry of the ramp.
2. The bridge after cleaning.

1. The bold structure in its original state, serving today's traffic.
2. General view of the bridge.

IM 96. Ponte di Porta Cappucina

River: Tronto, road: Via Salaria, location: Ascoli Piceno, province: Ascoli Piceno. Length: 62 m, width: 6.32 m, height: 24.8 m, span: 22.2 m, material: mostly travertine, ratio of voussoir to radius: 1/11.1 m = 0.090. Built under Augustus, time of construction: 1st c. BC. Visited: April 1989.

This bridge is also called Ponte Romano or Ponte di Solesta. The construction of its large span, slender arch and high embankment will have been influenced by the great Narni bridge. It is grounded on bedrock and 20 m high abutment walls secure the big arch. The arch with 57 voussoirs 60 x 100 cm is accentuated by a small moulding, as are its springing from the abutments and the level of the pavement with the drainholes. The masonry is carefully finished, giving the structure a refined appearance and making this bridge an outstanding example of the Roman's art of bridge building . Ascoli Piceno lies on a plateau surrounded by the rivers Tronto and Castellano. It is reached from the south via this bridge, which is easily taking the modern traffic.

Ref.: AB 1943, p. 677; Galli and Viccei 1940, p. 87; Gazzola 1969, p. 65.

IM 97. Ponte Cecco

1, 2. Two views of the bridge.

River: Castellano, road: Via Salaria, location: Ascoli Piceno, province: Ascoli Piceno. Length: 45 m, width: 6.32 m, height: 24.8 m, spans: 14.5 + 7.15 m, material: mostly travertine, ratio of voussoir to radius: 79/725 cm = 0.108. Built under Augustus, time of construction: 1st c. BC. Visited: April 1989.

This bridge provides another access to the town Ascoli Piceno, which is protected by the confluence of the Tronto and Castellano rivers. The bridge has a daring touch – similar to IM 97 – due to the narrow and deep ravine. The main arch has 51, about 45 x 80 cm voussoirs of travertine. It was destroyed during World War II and rebuilt to its former state with a small commemorative middle gate. It is hardly accessible, because on one side it runs against the town's fortification, on the other against the steep embankment.

Ref.: Gazzola 1969, p. 66.

1. Fully exposed arch.
2. The hanging keystone.

IM 125. Arci San Vidam

Road: Branch of Via Appia ?, location: near Sermoneta, province: Latium. Length: 13.75 m, width: 3.5 m, spans: 5.2 + 4.45 m, material: limestone, ratio of voussoir to radius: 90/260 cm = 0.34. Visited: July 1992.

The two arches are all that remains of what seems to have been a causeway across the Pontine Marshes. A connecting road exists no more. The remains demonstrate the stability of well built arches, even after having lost their abutments. They consist of 19 or 13 large voussoirs 50 x 90 cm and up to 110 cm deep. One arch is about to collapse, as the keystone hangs precariously.

Even Gazzola does not mention the structure.

IM 127. Ponte dell'Abbadia

River: Fiora, location: Vulci near Canino, province: Latium. Width: 3 m, height: 35 m, span: 25 m, material: Tuff, limestone, ratio of voussoir to radius: 1.5/12.5 m = 0.1200, time of construction: 1st c. BC. Visited: June1985.

The bridge crosses in one bold arch the Fiora river. The abutments of reddish tuff and the spandrel walls of grey tuff are original, the humped limestone arch (57 voussoirs 69 x 150 cm) has been rebuilt in limestone. Consoles for the falsework remain. Gazzola claims a span of 35 m, which is wrong, since the semi-circle of voussoirs adds up to »only« 25 m. Long ago the bridge served as aqueduct, the leaking water producing picturesque stalactites. The singular structure has inspired rapturous accounts by British travelers, such as »an overwhelmingly romantic place« or »this wonderful structure straddles the rocky chasm like a giant«, or »the bridge appears like a lonely rainbow against the sky, in a perfect elegance«. On the east side of the bridge is a medieval fortress, which marked the border of the Papal States.

Ref.: Blake 1974, p. 210; Gazzola 1963, p. 3.

1. View of the pavement.
2. General view with fortress.
3. The nearly dry Fiora stream crossed by the bridge.

1. General view of the arch.
2. Cross view of the structure, showing 1.8 m wide parapets and the 3 m wide pavement with its core of opus caementitium.

F 4. Pont d'Ambroix

River: Vidourle, road: Via Domitiana, location: Gallargues, department: Gard. Width: 6.6 m, height: 7 m, spans: 1 x 10 m remains, material: limestone, ratio of voussoirs to radius: 65/500 cm = 0.1300, pier width: 2.7 m. Built under Augustus, time of construction: 1st c. AD. Visited: June 1991.

The bridge was near the ancient Oppidum Ambrossum. Only one arch remains of five arches to cross a river bed of about 150 m. It may have had up to eleven arches, before the river narrowed to its present bed. The semicircular vault consists of four carefully crafted arches with infill in between. The piers have square floodwater openings. Impost corbels served as support of the falsework. A strong molding accentuates the top of the piers.

Ref.: Chevallier 1979, p. 211; Fasti 1973/74; Fiches 1970, p. 142; Gazzola 1969, p. 124; Rivet 1988.

F 6. Pont St.Thibery

River: Hérault, road: Via Domitia, location: St.Thibery, department: Hérault. Length: 150 m, width: 4 m, height: 4 m, spans: four of 9.9 to 11.1 m (out of nine), material: local volcanic stone, ratio of voussoirs to radius: 75/495 cm = 0.1515, pier width: 4.7–5.5 m. Built under Augustus, time of construction: 1st c.AD. Visited: June 1991.

Four of nine arches remain at the western end. The segmental arches rest on wide piers with pronounced cutwater projections. A 16th-century flood destroyed most of the bridge. The flood-waters were particularly large (1.8 m wide by 1.7 m high) and are protected by the massive cutwater masonry, up to 5 m deep and now protected by reinforced concrete. The bridge is southeast of St.Thibery.

Ref.: Chevallier 1979, p. 211; Gazzola 1969, p. 126; Prade 1986, p. 211; Rivet 1988.

1. General view from north.
2, 4. Details of the exposed bridge structure showing the combination of cut stone and concrete.
3. One of the graceful arches.

1. Close-up of an arch.
2. General view of the bridge.

F 9. Bridge at Boissezon

River: Bénovie, road: Via Domitia, location: Bénovie, department: Hérault. Length: 48.9 m, width: 3.55 m, height: 9.5 m, spans: 6 + 8.7 + 9.4 + 8.7 + 6 m, material: local limestone, ratio of voussoirs to radius: 48/470 cm = 0.1021, pier width: 2.5 m. Built under Agrippa, time of construction: 1st c. BC. Visited: June 1991.

This bridge captivates by the symmetrical arrangement of arches with increasing spans and identical piers. It is very well built and was widened by 3 m to 6.50 m by the addition of arches in the original manner. It rises slightly to the middle arch, the keystones are projecting for emphasis. Voussoirs are 45 x 75 x 175 cm deep. Falsework supports are about 1 m higher than the impost level to reduce the amount of shuttering. The bridge is still in full use.

Ref.: Gazzola 1969, p. 127.

F 10. Pont Julien

1. Elevation of the bridge.
2. The rise to the central arch, metal hunters' damage to the vault.
3. View of the principal arch.

River: Coulon, road: Via Julia Augusta, location: 8 km west of Apt, department: Vaucluse. Length: 115.5 m, width: 5.9 m, height: 14 m, spans: 10.5 + 16.2 + 10.5 m, material: limestone, ratio of v/r: 85/810 cm = 0.1049, 75/525 cm = 0.1428, pier width: 3.8m. Built under Augustus, time of construction: c. AD 10. Visited: June 1991.

The name of the bridge seems to derive from Caesar's dynasty the Julians. The site was carefully chosen, down or upstream no better place could be found. The bedrock was carefully prepared to receive the foundations of the piers. This way the bridge survived various floods that destroyed bridges in the vicinity, such as the Pont des Beaumettes. A 49 m long northern and a 20 m long southern ramp connect with the bridge, which rises to the top of the central arch. Large floodwater openings (1.5 x 3.1 m) pierce the spandrels. At pavement level a triangular molding projects 25 cm. A parapet made of blocks 60 x 80 x 145 cm carries a metal railing. The net road width is 4.3 m.

In the masonry of the vaulting we notice many deep holes. Most likely the stones were linked by metal anchors, which prompted metal thieves to chisel them out.

Ref.: Chevallier 1979, p. 211; Emerson and Gromort 1925; Gazzola 1969, p. 127; Monkewitz 1982, p. 29; Prade 1986, p. 380; H. Schreiber 1985.

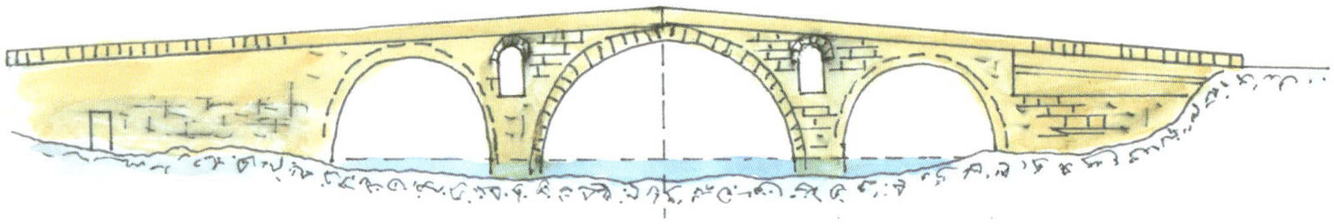

1. View of the bridge.

F 11. Pont sur l'Ouvèze

River: Ouvèze, road: off Via Domitia, location: Vaison la Romaine, department: Vaucluse. Length: 18 m, width: 9 m, height: 13 m, span: 16.2 m, material: granite, ratio of voussoirs to radius: 1/8.1 m = 0.1234. Built under Agrippa, time of construction: 1st c. AD. Visited: June 1991.

This single-arch bridge is fitted between the rock faces of the ravine at its narrowest point. The vault is slightly parabolic, similar to IM 04, IM 90 and IN 35. It consists of five arches connected by metal ties and starts at 3 m above water level. The structure was damaged during World War II and rebuilt in 1954. It is in full service.

Ref.: Chevallier 1979, p. 164; Gazzola 1969, p. 128; Prade 1986, p. 386; Rivet 1988.

F 13. Pont Flavien

1. The abandoned bridge in the fields.
2. Corinthian pilaster with entablature.
3. Bridge and commemorative gates.
4. Present state of the pavement.
5. The author measuring the depth of the rut.

River: Touloubre, road: Via Julia Augusta, location: St. Chamas, department: Bouches-du-Rhône. Length: 25.4 m, width: 6 m, height: 7.8 m, span: 13.7 m, material: limestone, ratio of voussoirs to radius: 1/7 m = 0.0142. Built under Donnius Flavius, time of construction: 12 BC. Visited: June 1991.

This small but very solid bridge received two elaborate 6.7 m high and 7.6 m wide commemorative gates, with full entablature and Corinthian pilasters. Even four crowning lions are still in place. The well-to-do builder commissioned them in his testament. The arch has exceptionally heavy voussoirs: 55 cm wide, 1 m high and 2 m deep, adding up to 1 m^3 or 2.7 tons each. In 1763 the bridge was thoroughly renovated.

In contrast to the well preserved gates and stone work, the pavement has been severely worn down. Up to 65 cm pavement thickness have been lost, exposing the crest of the vault and leaving only 35 cm for the voussoirs. Contrary to the rules of statics it is perfectly stable..

Ref.: Chevallier 1979, p. 211; Emerson and Gromort 1925; Fasti 1967, 5702; Gazzola 1969, p. 129; Prade 1986, p. 109.

1. The bridge.
2. A parapet block with the groove for tongue connection.

F 19. Bridge at St.Gabriel

River: local, road: Via Domitia, location: St.Gabriel on D33, department: Bouches-du-Rhône. Length: 12.25 m, width: 4.3 m, height: 4.5 m, span: 3.25 m, material: sandstone, ratio of voussoirs to radius: 50/165 cm = 0.3030. Visited: June 1991.

This modest bridge is a sample of the countless secondary structures needed for a road. Although grown over, it is in good condition. 4.5 m long wing walls reduce the road width from 6 to the 3.3 m of the bridge. The parapets consist of blocks of 50 x 50 x 200 cm with tongue and groove connections. On 7 April 1939 the structure made it into the French list of *Monuments Historiques*. 7 km south of Tarascon it is 100 meters from the Romanesque chapel St.Gabriel.

Ref.: Chevallier 1979, p. 211; Gazzola 1969, p. 178.

F 25. Bridge at Viviers

River: Escoutay, road: Via Antoninus Pius, location: Viviers, department: Ardèche. Length: 107.15 m, width: 4.25 m, height: 7 m, spans: eleven of 5.5 to 13 m, material: limestone, ratio of v/r: 55/650 cm = 0.085, pier width: 1.55–4.95 m. Built under Antoninus Pius, time of construction: 2nd c. AD. Visited: June 1991.

A very wide but shallow river had to be crossed by the road running along the Rhone River. The bridge has been built with material readily available in the river bed, the voussoirs are slightly trimmed and mortar has been used in abundance. This indicates the rather late date of its original construction, block sizes are small, voussoirs 32–35 x 40–55 cm only, the parapet walls are of rough irregular stones from the river (*opus incertum*). Ashlar masonry was no more used during the time of Antoninus Pius. In 1767/68 the crossing was fully renovated, including reconstruction of wide arches in parabolic shape. The flow opening is 71%.

Ref.: Chevallier 1979, p..171; Prade 1986, p..71.

1. Elevation of the bridge.
2. Bridge and riverbed.
3. Parabolic arch of later construction
4. Upstream side of bridge.
5. The romantic setting of the bridge.

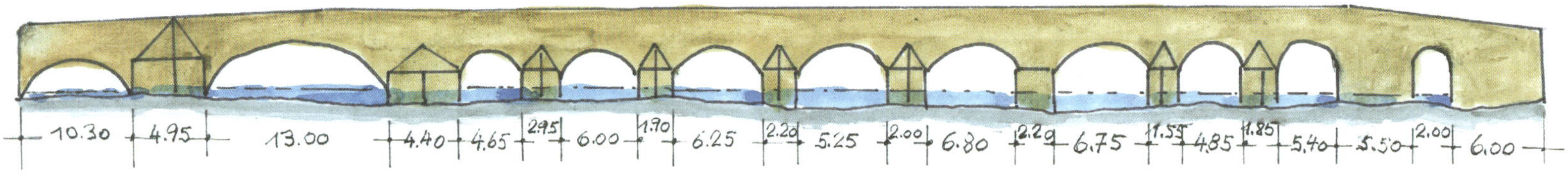

1–3. Detailed views.
4. Elevation of the aqueduct.
5. General view.

F 26. Pont du Gard

River: Gardon, road: local, location: Pont du Gard, department: Gard. Length: 275 m, width: 6 m height: 49 m, spans: see below, material: local limestone, ratio of voussoirs to radius: 145/960 cm = 0.1510. Built under Agrippa / Claudius, time of construction: 1st c. AD. Visited: June 1991.

This world-renowned structure served as an aqueduct, supplying Nice with water from 50 km away. The first level left a 70 cm footpath on either side of the next arcade, the road has been added in 1746.

There are three tiers of arcades:

1. Lower level of 6 arches with 19–24.4 m span, 142 m long, 6.3 m wide, 22 m high.
2. Middle level of 11 arches with 15.5–19.2 m span, 242 m long, 4.5 m wide, 19.6 high.
3. Top level of 35 arches of 4.8 m span, 275 m long, 3.06 m wide, 7.4 m high.

It is the tallest Roman bridge construction, surpassing Alcántara (42 m) or Narni (30 m). No mortar was used, except for lining the 1.25 m wide and 1.8 m high water channel on top.

The reader is kindly referred to the extensive literature regarding this famous structure.

Ref.: Brogan 1953, p. 82; De Camp 1977, p. 195; Emerson and Gromort 1925, p. 25; Frontinus Ges. 1988, p. 207; Grenier 1931/4, p. 88; Nikolajev 1967, p. 38; Prade 1986, p. 179.

F 27. Pont Vieux at Béziers

1, 2. Various views of the bridge.
3. Aerial view of Béziers with the Pont Vieux.

River: Orb, road: Via Domitia, location: Béziers, department: Hérault. Length: 241 m, width: 5.45 m, height: variable, spans: ten of up to 18.5 m, material: limestone, ratio of voussoirs to radius: 70/925 cm = 0.1060, time of construction: 12th c. AD. Visited: June 1991.

The Via Domitia is the oldest Roman road in Gaul, leading on to Spain. It played an important role during the campaigns of Caesar, Pompey and Augustus. Although the existing bridge was built in the 12th-century Romanesque period, it revives Roman construction excellence. Most likely it rests upon Roman pile foundations. Much building material stems from Roman times. The arches are segmental, the largest having 93 voussoirs of 30 x 50 to 70 cm. There are triangular cutwaters and floodgates. Charles VI and Louis XI (15th century) talk of a very ancient, sumptuous bridge. In the 14th, 15th and 16th centuries it was renovated.

Ref.: Chevallier 1979, p. 78.

1. General view.
2. Roman impost blocks and medieval abutments.

F 28. Bridge at Ganagobie

River: Buès, road: RN 96, location: Ganagobie, department: Alpes-de-Haute-Provence.
Length: 28.5 m, width: 5 m, height: 8.35 m, span: 7.9 m, material: local stone, ratio of voussoirs to radius: 60/395 m = 0.1519. Built under Hadrian, time of construction: 2nd c. AD. Visited: June 1991.

Already Strabo mentioned the tributaries of the Durance River, of which the Buès is one. Similar to contemporary bridges in Liguria, consisted the arch of two superposed ranges of voussoirs. During medieval restoration works most of the second range was omitted. The rather small scale medieval masonry contrasts with the massive impost blocks of Roman times. In 1962 the bridge was re-established as Roman by Guy Barruel and Pierre Martel. The bridge is situated 22 km east of Monosque at the branching from RN 96 to Lurs and the famous Ganagobie Monastery.

Ref.: Baruol in Gallia XXI, 1963, p. 314; Chevallier 1979, Fasti 1963/64, 9012; Pervin 1978; Prade 1985, p. 54; Rivet 1988

SP 09. Puente Romano over the Rio Guadiana (Spain)

River: Guadiana (ex Anas), road: off Via Augusta, location: Mérida, province: Badajoz. Length: 792 m, width: 7.9 m, height: 11.7 m, spans: 60 from 6.8 to 11.2 m, material: granite, north ramp: 75/560 cm = 0.1339, south ramp: 80/300 cm = 0.2666, pier width: 4.75 m. Built under Trajan, time of construction: 1st to 2nd c. AD. Visited: April 1990.

This longest Roman bridge consists of a north ramp with 12 arches, two of them hidden. The main bridge has 26 arches and the south ramp 22 arches. It is in very good condition and in full use, but the construction history is involved, since the river kept changing its course in the wide alluvial plain. It is believed that Augustus started a bridge in 25 BC, which later was continuously enlarged, to accommodate flooding. Casado writes of three bridges, each an addition the previous one. The net flow width of the bridge is 64 % of its length. The Visigoth king Hervigius ordered repairs in 686, a major restoration followed in 1610. From 1822 until 1956 flood damaged piers and arches were repaired.

Augusta Emerida – today's Mérida – was a city for veterans with all the trappings of a proper Roman town, such as temple, forum, theatre, fortification. The bridge is very impressive due to its excellent condition and harmonious proportions. An inscription names Trajan as its builder: *Imp Caesari Div Nervae Traiano Aug Germ Dacico Ponti Max Trib Potes VIII Imp V Cos VPP*.

Ref.: Alvarez-Martinez 1983, p. 93; Casado 1972, p. 11; Gazzola 1969, p. 121; Merckel 1902, p. 301; Schreiber 1985, p. 117.

1. Arches, piers and rounded cutwaters.
2. Detail of a pier with floodway.

1. General view.
2. Exact semicircles reflect in the water. A cornice at pavement level is cut by water spouts.

SP 12. Albarregas Bridge (Spain)

River: Albarregas, road: Via Augusta, location: Mérida, province: Bajadoz. Length: 125 m, width: 7 m, height: 4.5 m, spans: 4 x 5.2 m, material: granite, ratio of voussoirs to radius: 50/260 cm = 0.1923, pier width: 2.5 m. Built under Trajan / Hadrian, time of construction: 2nd c. AD. Visited: April 1990.

This bridge crosses a confluence of the Guadiana in the northern part of Mérida. Four identical arches in accentuated masonry give a strong and harmonious impression. They are in the original condition, each with 21 voussoirs of 40 x 50 x 60 cm. The ramps, often repaired, are adding nearly100 m to the length of the bridge. They have several floodwaters. The parapet is designed with slightly projecting balusters and below the drains at paving level is a strong molding. Next to the bridge is the lofty Los Milagros aqueduct. The Via de la Plata from Lisbon over Merida to the interior later followed this route.

Ref.: Gazzola 1969, p. 137; Grewe 1993, p..244; Merckel 1902, p..301.

SP 15. Bridge over the Ambroz (Spain)

1. Detail of masonry and parapet.
2. General view of the bridge.

River: Ambroz, road: Via de la Plata, location: Ventas de Cáparra, province: Cáceres. Length: 64.7 m, width: 7.45 m, height: 5.5 m, spans: 5.4 + 2 x 8.9 + 5.4 m, material: granite, ratio of v/r: 62–85/445 cm = 0.1393–0.1910, pier width: 4 m. Visited: April 1990.

Also this bridge is in excellent condition and in active use. A small molding at pavement level sits below the neat, small scale masonry of the parapet, which is articulated by pedestals at about 4 m intervals. The large arches consist of 35 voussoirs, 40 x 62-85 x 60-100 cm.

The bridge is 17 km north of Villar de Plansencia in the direction to Salamanca, then 10 km west towards Ahigal. Nearby was the Roman town Caparra.

Ref.: Casado 1972, p. 186; Gazzalo 1969, p. 139.

1. The fateful sign on the highway.
2. Partial view of the bridge.

SP 22. Brodalo bridge (Spain)

River: Brodalo, road: Medina to Trujillo, location: Miajadas, province: Bajadoz. Width: 7.45 m, height: 5.5 m, spans: 3 x 10 + 8 m. Visited: April 1990.

This is the bridge which prompted the author's decision to study the Roman work. A sign directed him off the modern highway to the Roman structure. The bridge is so well preserved that it is most likely a later reconstruction on Roman foundations. It is 30 km north of Mérida, near the town Miajadas.

SP 29. Trajan's Bridge at Alcántara (Spain)

River: Tejo, road: off Via Augusta, location: Alcántara, province: Cáceres. Length: 194 m, width: 8 m, height: 48 m, spans: 18.5 + 24.3 + 27.3 +28.5 + 24.3 + 18.5 m, material: granite, ratio of v/r: 62-85/445 cm = 0.1393-0.1910, pier width: 8.5 m. Built by C. Julius Lacer, time of construction: AD 104. Visited: April 1990.

Curiously, this huge bridge is not part of the major Roman roads in Spain. A small temple at the east side of the bridge contains an inscription, telling us that C. Julius Lacer built this bridge to last in eternity – a wish which has been fulfilled, since it is well maintained and serves modern traffic. There is a 14 m high commemorative arch across the middle of the bridge, with a dedication to Emperor Trajan in the 5th year of his rule, indicating the date AD 104. due to its height the bridge has very large piers, 8.3 to 9 m wide and up to 25 m deep. Floods and wars have repeatedly damaged the structure, but it has been always well repaired. Today a huge barrage close by prevents flooding.

Ref.: Blanco 1971, pp. 142, 154, 161; Callecho Serrano 1970, p. 213; Casado 1972, pp. 41, 141, 154, 161; Fasti 1971/72, 8830; Gazzola 1969, p. 133; Merckel 1902, p. 299; Pekáry 1968, p. 125; Wiseman 1956; Wölfel 1990.

1. The bridge seen downstream.
2. Granite masonry for eternity.
3. Temple at the eastern end.
4. Huge piers and projecting masonry strengthens the bridge.

1.
2. The reconstructed arch of the bridge.

SP 30. Bridge at Guijo de Granadilla (Spain)

River: Alagón, road: Coria to Calcadilla, location: Guijo, province: Cáceres. Width: 5.6 m, height: 8 m, span: 20 m, material: granite. Visited: April 1990.

This bridge consisted of a single arch and has been submerged by a reservoir. It was reconstructed above water level as a historical monument – a rather unique reverence to Roman ingenuity.

Ref.: Casado 1972, p. 90; Gazzola 1969, p. 189; Merckel 1902, p. 302.

SP 41. Bridge over the Tamega River (Portugal)

River: Tamega, location: Chaves. Length: 116 m, width: 6.8 m, spans: 18 (today 9) x 8.95 m, pier width: 2 m. Built under the Flavians, time of construction: 2nd c. AD. Visited: April 1990.

An inscription dedicated this bridge to Vespasian, Titus and Domitian. The town has encroached upon the bridge, obscuring the original number of arches and length. Gazzola notes 18 arches, Wiseman 16. Two piers have projections with columns, upstream there are round cut waters.

Ref.: Casado 1972, p. 362; Gazzola 1969, p. 119; Merckel 1902, p. 301; Wiseman 1956, p. 143.

1. General view of the bridge.
2. The bridge entering the town.

1. Fully serviceable till today.
2. General view of the bridge.
3. An elegant structure.

SP 42. Bridge at Villa Formosa (Portugal)

River: Seda, road: Via Augusta, location: Alter do Chao, province: Ribatejo. Length: 115.3 m, width: 6.75 m, height: 8.9 m, spans: 6 x 8.8 m, material: granite, ratio of v/r: 100/440 cm = 0.2273, pier width: 3 m. Built during imperial time, time of construction: 1st–2nd c. AD
Visited: April 1990.

The harmonious design with six identical arches is accentuated by its embossed masonry. The impost levels are emphasized by a molding at the top of each pier. The bold semicircular arches have 33 voussoirs each (42 x 100 x 80 to 90 cm deep). A couple of small floodways and a crowning projection at road level enliven the structure further. Ramps of 28.5 and 19 m complete the bridge. The original pavement carries today's traffic. In 1950 the bridge was restored, keeping the original parapets..

Ref.: Fasti 1948, p. 37; Gazzola 1969, p. 122; Pellati 1940, p. 62.

SP 43. Bridge over the Elja River (Portugal)

1. General view of the bridge.
2, 3. Close-up of the masonry.

River: Elja, road: Alcántara to Castello Bianca, location: Segura, province: Alentejo. Length: 84 m, width: 6.5 m, height: 15 m, spans: 2 x 9.4 + 12.4 + 2 x 7.9 m, material: granite, ratio of v/r: 70/620 cm = 0.1129 (large arch), pier width: 3 m. Built under Trajan, time of construction: 1st c. AD. Visited: April 1990.

Since the imposts of all arches are at the same level, the widest arch determines the height of the road and necessitates additional masonry above the smaller arches. Since all voussoirs are of the same size (70 x 40 cm) and the arches spring from the same level, the bridge has an extraordinarily harmonious appearance. The biggest arch has 49 voussoirs, the next smaller 37 and the smallest 31. It looks like the builders ran out of granite blocks, which are beautifully bossed but stop arbitrarily in the course of construction. The bridge was completed with irregular small scale masonry that contrasts with the previous work. It anticipates the slightly bigger Alcantara Bridge by Julius Lacer.

Ref.: Casado 1972, p. 145, 161, 184; Gazzalo 1969, p. 137.

1. The main arch from the south.
2. Central pier.
3. Old view of the bridge from Merckel, 1902.
4. General view of the bridge.

T 01. Mouslouk Bridge

River: Selinus, road: Adramyttium-Thyatira, location: Bergama, province: Izmir, West-Turkey. Length: 38 m, width: 5 m, height: 10 m, spans: 9.1 + 12.6 m, material: trachyte, concrete, ratio of voussoirs to radius: 60/4.5 = 0.1319, 70/6.3 = 0.1111. Built under Hadrian, time of construction: 2nd c. AD. Visited: July 1990.

This bridge over the Selinus River, today called Bergama Cayi, is within the city of Bergama, ancient Pergamum. It merges with the buildings of the town and is properly maintained, although the pier in the middle is seriously deteriorated. The highly resistant trachyte facing and the concrete core assure a solid structure, which is still in full use. The big arch describes 90 % of a semicircle and has 51 voussoirs of 35 x 70 cm, the smaller arch has the same number of voussoirs, but of 25 x 60 cm size; they are up to 120 cm deep. Seven courses above the foundation projecting consoles were supporting the falsework.

Ref.: Gazzola 1963, p. 147; Merckel 1902, p. 303; Tunc 1978, p. 33.

T 02. Twin Tunnel at Pergamum

River: Selinus, road: Adramyttium-Thyatira, location: Bergama, province: Pergamon. Length: 196 m, width: 33 m, height: 7.5 m, spans: 2 x 9 m, material: trachyte, concrete, ratio of voussoirs to radius: 77/450 cm = 0.1711, pier width: 2.5 m. Built under Hadrian, time of construction: 2nd c. AD. Visited: July 1990.

Although having the aspect of a tunnel, this structure is really a very wide bridge of two parallel vaults. Its has been built to create a 110 by 260 m square in front of the venerable »Red Basilica«, (Turkish Kizil Avli), possibly a sanctuary of the Egyptian triad of Isis, Serapis and Harpocrates. Every 7 m the »tunnel« has neatly cut stone arches of a slight horseshoe shape, guiding the form-work of the concrete vault in between. They rest on a 3 m high and 2.5 m wide wall of heavy ashlar masonry. The downside ends of the tunnels are staggered by 13 m, reducing the length of the shorter tunnel to 183 m. The tunnel allows a flow of 720 m^3/sec, sufficient for major floods. Today the town covers the structure with up to 5-storey buildings and its modern traffic. The structure is 40 m downstream from the Mouslonk Bridge (T01).

Ref.: Grewe, Özis, Baykan, Atalay 1949, p. 348; Merckel 1902, p. 303; Tunc 1978, p. 33.

1. View of the tunnel entrance from the north.
2. Tunnel exit.
3. Plan of the square in front of the »Red Basilica« with the tunnel underneath (after Greve).
4. Detail of entry pier.

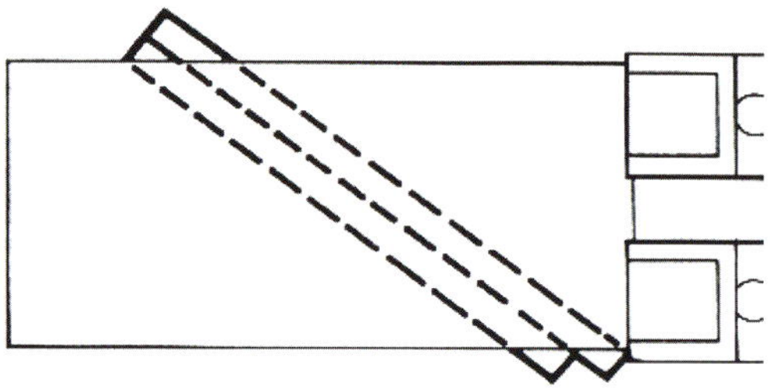

1. General view of the bridge.
2. Today's pedestrian crossing.
3. Detail of pie.r
4. Sample of author's site measuring drawing.
5. Close-up view of bridge.

T 05. Marsias bridge

River: Marsias, road: Stratonikaia-Gergas, location: Ince-Kemer, province: Aydin. Length: 77 m, width: 2.15 m, height: 15 m, spans: 16 + 2 x 13.3 + 9.8 + 6.2 m, material: granite. Visited: Sept. 1987.

Although only good for pedestrians or riders, this structure is an outstanding example of Roman engineering. Most likely it was an aqueduct serving the town of Gergas. The bridge rests on solid rock, assuring the stability of the structure throughout millennia. 200 m further another small bridge crosses a confluence of the Marsias River, foundations of three arches can be seen. Pierced tubular stones can be found in the vicinity, indicating the past existence of a water conduit.

Ref.: Bean 1966; Dörner 1984, p. 147.

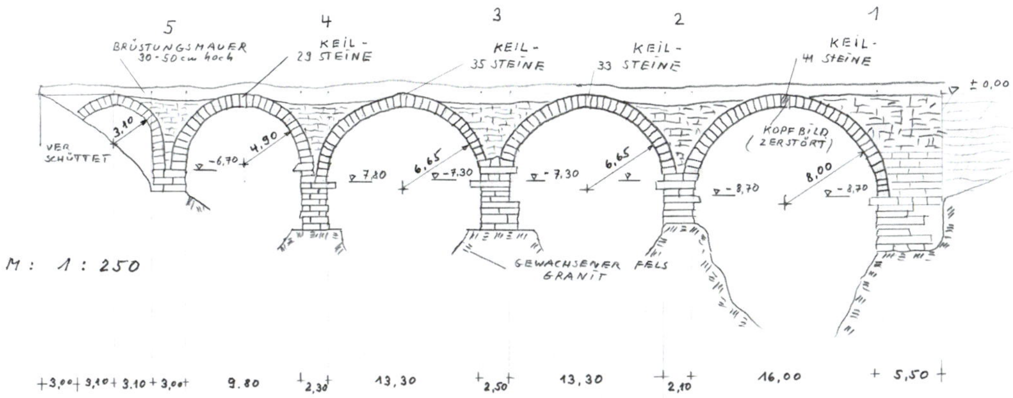

T 10. Aqueduct-Bridge at Troy

River: Thymbrios, location: Kemer Dere (Troy), province: Çanakkale. Length: 70 m, width: 1.4 m, height: 25 m, span: 18 m, material: limestone, ratio of voussoirs to radius: 80/900 cm = 0.0888. Built under Herodes Atticus, time of construction: 2nd c. AD. Visited: August 2000.

Although it is an aqueduct this structure is included for its elegance. It was built to supply Troy with water. The water channel was later raised with rubble masonry and small arches. Gibbon recounts that Herod Atticus as prefect received from Hadrian a huge sum for the construction of a water conduit to Troy. When the construction costs turned out to be double the estimate, his father Atticus paid the extra out of his pocket. In Byzantine times the aqueduct level was raised by about 2 m, adding rough stone masonry pierced by intermittent arches.

Ref.: Dörpfeld 1968, p. 241; Gibbon 1892.

1–3. Views of the aqueduct.

1. Pier projection with decorative molding.
2. Additional structures at the east end probably were harbor facilities.
3. Exedra at the eastern bridgehead.
4. The bridge in the fields.
5. A segmental double-arch.

T 12. Bridge over the Sangario

River: Sangario (Sakarya), road: Adramyttium-Thyatira, location: Adapazari, province: Pergamon. Length: 429 m, width: 10.3 m, height: 9.6 m, spans: 8 x 22.4 m, material: limestone, ratio of voussoirs to radius: 80/1120 cm = 0.0714, pier width: 7.6 m. Built under Justinian, time of construction: 561 AD. Visited: July 1990.

Today the bridge runs across fields, since the river Sakaria has shifted. Segmental arches are used to keep the height down. The two-tier voussoirs together with the segmental circle shape are typical for the time of construction during late imperial times. Semicircular pier projections divert the water from the wide supports. About 180 m of the length of the bridge consist of ramps. Froriep's study of water management in Bithynia suggests that the bridge crossed the corrected flow of the Melas River. At the eastern bridgehead is an exedra, emptied of its contents.

Ref.: Cüppers 1969, p. 165; Froriep 1986, p. 39; Gazzola 1963, p. 174; Texier 1836, p. 55 ff.; Tunc 1978, p. 156; Whitby 1985.

T 14. Bridge at Beskonak

River: Eurymedon, road: Aspendos-Selge, location: Aspendos, province: Antalya. Length: 20 m, width: 3.4 m, height: 30 m, span: 13.6 m, material: local stone, ratio of voussoirs to radius: 70/680 cm = 0.1029, time of construction: 2nd c. AD. Visited: September 1987.

Again the rock embankments give the bridge with its daring arch (61 voussoirs of 33 x 70 cm) an excellent stability. It still is serving perfectly today's traffic. The Eurymedon – Turkish Köprü Cay – is today a rafting destination.

To reach the bridge turn northward from the coastal road at the height of Aspendos in the direction of Tasagit–Beskonal–Selge. The large antique town of Selge is 10 km behind the bridge. It was risky to take measurements since the bridge has no balustrade.

Ref.: Bean 1968, p. 139.

1. Record of measurement – a challenge for the author.
2. Close-up of the bridge.
3. Modern traffic on a 2000 years old bridge.

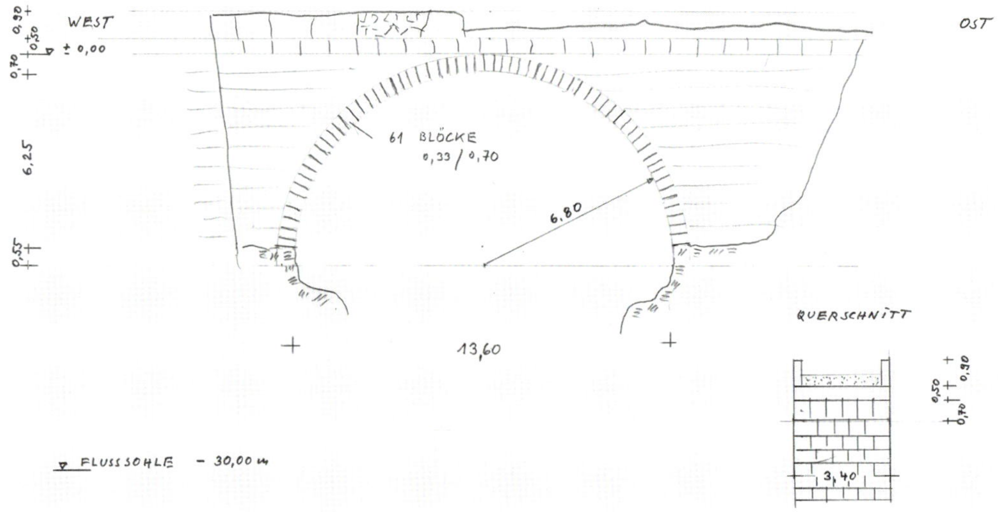

1. The alluvial plain.
2. Paving.
3. Segmental brick arch.

T 16. Bridge at Limyra

River: Alakir Cay, road: to Kumluca, location: Limyra-Toruncova, province: Finike. Length: 360 m, width: 3.7 m, height: 4 m, spans: 28 x 9.5 to 12.9 m, material: local stone, brick, ratio of v/r: 80/640 cm = 0.1250, pier width: 2.1 m, time of construction: 6th c. AD ?. Visited: September 1987.

Today this bridge is in an alluvial plain, the river shifted eastward. Together with the bridge of San Lorenzo in Padua, the bridge at Kemer in the Xanthos valley, Justinian's bridge at Adapazari and the Alconétar bridge in Spain, it is one of the few Roman bridges with segmental arches. In this case the flat (140°) arches (span to rise = 0.24) were built with two ranges of flat brick (39/50/5 cm). With a 2 cm mortar joint the superposed brick give an arch height of 80 cm. The pavement consists of 10 to 20 cm thick stone slabs of up to one meter diameter. Originally having 27 arches, the 27th arch has been rebuilt with two smaller arches, giving a total of 28 arches. According to today's standards the bridge could carry a 30 ton truck plus 500 kp/m^2 live load. As always, the bridge was overdesigned for antique traffic. Much of the bridge is submerged in the plain, which is used for agriculture. Its small width indicates that it served mostly pedestrian and animal traffic.

Ref.: Bean 1978; Fasti 1977/78; Wurster and Ganzert, 1978.

T 35. Chabinas bridge

1. The bridge today.
2. Drawing of the bridge by Colin O'Connor.

River: Chabinas, road: Samosata-Melitene, location: Killisik/Eski Kahtâ, province: Eastern Turkey. Length: 116.8 m, width: 7.56 m, height: 17.1 m, span: 34.2 m, material: local stone, ratio of voussoirs to radius: 70/680 cm = 0.1029. Built under Severus, time of construction: 2nd c. AD. Visited: September 1987.

This bridge over the Chabinas – now Cendere Cay – is part of the road from Samosata to Metilene. It was built by the Legio XVI Flavia Firma commanded by Lucius Marius Perpetuus at the time of Septimus Severus in preparation of a campaign against the Parthians. Four cities of the Commagene: Samosata, Doliche, Perrhe and Germanicia shared the construction costs. It has the second largest span after the bridge of the pont St. Martin in the Aosta valley (35.6 m) or by Augustus at Narni (32.1 m). Its vault is a 70 % segment of a full semicircular arch. Typically for the late Empire and for Byzantine times it is approaching a parabolic shape with its excellent structural performance. The voussoirs are 60 cm high and 187 cm deep.

Remarkable are four columns dedicated to the Emperor Septimus Severus, his wife Julia Domna, Emperor Caracalla and his brother Geta. Geta's column was dismantled after his murder by Caracalla. The bridge is 41 km east of Adiyaman, nearby the famous Nemrud Dâg, »throne of the gods«.

Ref.: Akurgal 1978, p. 347; Dörner 1984, p. 329, and 1987, p. 57; Gazzola 1963, p. 172; Jakopi 1936/37; Merckel 1902, p. 307; Tunc 1978, p. 43; Wagner 1975, p. 78.

NE 01. Bridge at Maameltein

This bridge has not been studied by the author. It is included as a special case and and a typical example of neglected Roman bridges. Nobody takes responsibility for its preservation. In this case a Lebanese committee (info@protect-lebaneseheritage.com) tries to promote awareness of the situation.

The road along the Mediterranean coast from Egypt to Syria crossed many torrents descending from the Lebanon Range. Among the surviving bridges is this one at Maameltein, 18 km north of Beirut (Berytus). It has a single 12 m arch; material has been dumped at the site.

Below it is the regular coastal road, above the modern expressway.

1. View of the bridge from the local road, standing on the Ottoman-time bridge. On the right the concrete core is exposed, otherwise neat ashlar masonry is in place. Above we see the modern expressway bridge. (Photo: David Corm.)

8. Glossary of technical terms

abutment	The end of a bridge where it joins the sides of a valley or natural ground
aggregate	A mixture of different mineral materials hardened by pressure or cement, or the gravel added to cement to form concrete
anathyrosis	The dressed edges of a squared block of building stone to ensure a good fit with little or no mortar
apse	A semicircular or polygonal recess in a church or temple
aqueduct	An artificial channel to carry water by gravity. A structure which carries such a channel across a valley
arch	A curved structure bridging an open space. It transfers loads to its supports primarily by compression
Archimedes' screw	A water pump which lifts water along a rotating inclined tubular screw
ashlar	A squared block of building stone
balustrade	A railing composed of a series of massive supports (balusters)
beam	A horizontal or inclined member transferring loads by virtue of its bending strength
bearing capacity	The load in kg per sq.cm which the existing ground can accommodate without destabilization
bed	The horizontal surface of supporting ground or masonry
bending moment	The force generated in the cross section of a member due to bending
block and tackle	A pair of pulley blocks with their ropes and attachments
caisson	A waterproof enclosure emptied of water and filled with concrete or masonry to support a pier below water level
cantilever	The free part of a horizontal member projecting beyond a support
capstan	A windlass with vertical axle
catenary	The curve assumed by a chain hanging freely between two supports at the same level
cement	In masonry construction a fine powder which hardens by combination with water
cement, hydraulic	A cement which hardens under water
centering	Temporary framing to support building materials during construction
chain dogs	A pair of metal hooks linked by a chain, used to lift a stone by engaging holes cut in two opposing faces of the block
cofferdam	A waterproof enclosure emptied of water, allowing workmen to build up the foundation of a pier below water level
compression	A force or stress which causes shortening of a member
concrete	A mixture of stones and mortar to be filled into formwork until it has hardened
console	A bracket-like member projecting from a vertical surface to serve as support
corbelling	An overlapping arrangement of stones, each horizontal course projecting beyond the one below
cornice	An ornamental projecting course, either horizontal or around the upper edge of an arch
cramp	A metal connection used to join stones in masonry construction
crank	An eccentric arm attached to an axle used for turning the axle
crown	The highest point of an arch
deep foundation	A foundation distributing the load downward into the ground
density	Mass per unit volume
durability	The ability of a material to resist water, frost, corrosion or wear
embankment	A mound or dike raised to hold back water or carry a roadway
extrados	The upper surface of an arch
facing	The application of stones of bricks as external finishing of a wall
false arch	An arch shape obtained by the corbelling method
falsework	Temporary support of a structure under construction
floodway	An opening above the pier of a bridge to allow the passage of flood water

foot (Roman)	About 0.294 m
force	An influence which causes a body to start moving, the acceleration being directly proportional to the magnitude of the force and inversely proportional to the mass of the body
formwork	Temporary support of concrete when it is first placed until it gains its strength
foundation	The base of a structure, designed to transmit the load of the structure to the ground
fracture	The process whereby a material fails by separation across a crack
friction	The resistance to the sliding motion of two bodies along their contact surface
funicular polygon	The shape of a flexible string when it is anchored at two points and loaded with a series of point loads
girder	A principal horizontal beam in a structure
header	A stone in ashlar masonry, with its length perpendicular to the wall face
hodometer	A distance measuring device
impost	The point where an arch rests on a wall or a column
intrados	The lower face of an arch
keystone	The central stone of a masonry arch
kg	Kilogramm, unit of mass = 2.205 lbm
kp	Kilopond, unit of force generated by 1 kg at standard acceleration
Lewis	A device used to lift a stone by engaging a specially cut hole in its upper face
lime	The material produced by heating limestone in a kiln to form calcium oxide (quicklime); or by addition of water to form calcium hydroxide (slaked lime)
line of pressure	The line representing the location of the resultant axial force in an arch
load	Another name for force. May be applied locally (point load) or distributed over an area (spread load)
mass	The amount of matter in a body
member	A distinct part of a structure, long rather than thick
moment	The product of a force and its perpendicular distance to a point or axis
mortar	A hardening mixture of lime or cement, sand and water
N	Newton, the force which gives a mass of one kilogram an acceleration of 1 m/sec/sec
opus caementitium	Masonry of crushed stone (caementa) or brick with mortar
– incertum	Masonry of irregular stone and mortar
– latericium	Brick masonry
– mixtum	Masonry of irregular stones at intervals squared off by a brick courses
– quadratum	Squared stone masonry
– reticulatum	A »network« masonry of small square stones laid diagonally
– rusticum	quarry faced masonry
– siliceum	Cyclopean masonry of irregular massive stones
– spicatum	Herring-bone masonry with inclined layers of shallow stones
– testaceum	Brick masonry
– vittatum	Masonry with irregular succession of stone and brick courses
parapet	A protective wall or barrier at the edge of a balcony, roof or bridge
pier	An intermediate vertical supporting structure in a bridge
pile	A member pointed at one end and driven into the ground to provide support to a structure
piledriver	A machine used to drive a pile
plug and feathers	A cut hole in the upper face of a stone receives a tapered rod (the plug) which is secured at the sides with flat metal sleeves (feathers). To split the stone the rod is driven in. If the hole is undercut, the plug widening and secured with feathers, it can be used to lift the block

plumb bob	A surveying device, consisting of a weight pointed at its base and hung by a cord
pozzolana	A variety of volcanic tuft or ash used in making hydraulic cement
pulley	A grooved wheel mounted on an axle, used to change the direction of a rope passed around the wheel
pulley block	A system of pulleys mounted on a single axle, or a pair of such blocks joined by a single rope passing around the pulleys.
reinforcement	In concrete work the addition of steel bars to take up excessive forces, mainly of tension
resultant	The diagonal force in a parallelogram of two different forces
rise	The height of the top of an arch above its springing
seam	A line formed by joining two layers
segmental	The adjective applied to an arch whose shape is only part of a semi-circle
shearing force	A force perpendicular to the length of a member
sheet piling	Boards driven next to each other into the ground to form a wall or sheet
soffit	The underside of a beam or an arch
span	The distance between adjacent supports
spandrel	The space between an arch and the horizontal level above its crown
spread footing	A foundation which spreads the load according to the bearing capacity of the ground
springer	The first stone of an arch resting at impost level
springing	The beginning of an arch
stretcher	In ashlar masonry a stone running parallel to the wall face
stress	force per unit area
string course	A projecting horizontal course in a wall
tension	An axial force or stress, pulling a material apart
tackle	A system of leverage using several pulleys or pulley blocks for multiplying power
torque	The turning power of a shaft
treadmill	A cage-like wheel with a horizontal axis, large enough to accommodate men walking within to make it turn
vault	An arched structure made of stones, concrete or bricks, bridging a space and supporting a floor, a roof or a roadway
voussoir	A wedge-shaped stone used with others to form an arch
waterwheel	A wheel with paddles turned by the flow of water
wedge	A tapered piece of wood or iron, driven into a gap to produce a large separating force
weight	The force applied to an object by gravitation
winch	A windlass turned by a crank
windlass	Device for raising weights attached to a cable or rope, by rolling up the cable over a horizontal axle

8. Bibliography

Adam, Jean-Pierre, *La construction romaine, matériaux et technique*, Paris, 1984.

Aimi, Amos, *Il ponte romano e la porta di S. Domino di Fidenza*, Parma, 1973.

Akurgal, Ekrem, *Ancient Civilizations and Ruins of Turkey: from Prehistoric Times until the End of the Roman Empire,* Istanbul, 1978.

Alberti, Leon Battista, *On the Art of Building in Ten Books,* Cambridge, MA, 1988.

Allain, Berthelot, and M. Gedell, »Pont gallo-romain et gué pavé à Argenton-sur-Creuse«*, Revue archéologique du centre de la France*, VI, 1963, p. 136–142.

Álvarez-Martinez, José Mariá, *El puento romano de Mérida*, Badajoz, 1983.

Amadei, Emma, *I ponti di Roma*, Rome, 1948.

Anderson, J. G. C., »The road system of eastern Asia Mino«*, Jounal of Hellenic Studies*, 17, 1897.

Andric, Ivo, *Bridge over the Drina,* Harvill, 1994.

Andric, Ivo, »Brücken«, Bihalji-Merin, Oto (ed.), *Brücken der Welt*, Lucerne, 1971.

Aschbach, Josef Ritter von, »Über Traians steinerne Donaubrücke«*, Mittheilungen der k.k. Central-Commission zur Erforschung und Erhaltung der Baudenkmäler*, III, 8, Vienna, 1858.

Ashby, Thomas, »Practical engineering in ancient Rome«*, Nature*, 116, 1925.

Ashby, Thomas, *The Roman Campagna in Classical Times*, London, 1927.

Ashby, Thomas, *The Aqueducts of Ancient Rome*, Oxford, 1935.

Ashby, Thomas, and Roland Arthur Lonsdale Fell, »The Via Flaminia«, *Journal of Roman studies*, XI, 1921.

Ashby, Thomas, and Robert Gardener, »The via Traiana«, *Papers of the British School at Rome,* 8, 1916.

Aubin, Hermann, »Die Rheinbrücken im Altertum und Mittelalter, eine Kriegs- und Wirtschaftsstudie«*, Rhein*, 7, 1937.

Augustus: *Res Gestae Divi Augusti: The Achievements of the Divine Augustus,* Oxford, 1967.

Ausonius, Decimus Magnus, *Mosella*, Stuttgart, 2000.

Avi-Yonah, Michael, *Encyclopedia of Archaelogical Excavations in the Holy Land*, Oxford, Jerusalem 1975/78.

Babelon, Ernest, René Gagnat, and Salomon Reinach, *Atlas archéologique de la Tunesie*, Paris, 1893.

Bagshave, R.W., *Roman Roads*, Aylesbury, 1979.

Baldoni, Daniela, »Il ponte romano di Savignano sul Rubicon«, *St. Romagnoli,* 30, 1979.

Ballance, M.H, »The Roman bridges of the Via Flaminia«, *Papers of the British School at Rome*, XIX, Rome, 1956.

Bammer, Anton, *Architektur und Gesellschaft in der Antike*, Vienna, Cologne, Graz ,1985.

Barradas, Lorenzo A., »Vias romanas das Regiôes de Chaves e Braganca«, *Revista de Guimarâes,* XVI, 1–2, 1956.

Baruol, Guy, »Le ponte romain de Ganagobie (Basse Alpes)«, *Gallia* , XXI, 1963/64.

Bean, George Ewart, *Aegean Turkey*, London, 1966.

Bean, George Ewart, *Turkey's Southern Shore, an Archeological Guide*, London, 1968.

Bean, George Ewart, *Beyond the Maeander*, London, 1971.

Bean, George Ewart, *Lycian Turkey*, London, 1978.

Bender, Helmut, »Römische Straßen und Straßenstationen«, *Kleine Schriften zur Kenntnis der römischen Besetzungsgeschichte Südwestdeutschlands,* 13, Stuttgart, 1975.

Bender, Helmut, »Römischer Reiseverkehr, Cursus Publicus und Privatreisen«, *ibid.*, 20, Stuttgart, 1978.

Benevolo, Leonardo, *Storia delle città*, Rome, 1975.

Besnier, P. T., »Entry Pons«, *Dictionaire Antique,* IV, 1905.

Bessac, J. C., M. Vacca-Goutoulli, and D. Lukas, *Im Schatten des Pont du Gard*, *Antike Welt,* 33, 2, 2002.

Bidwell, P. T., and N. Holbrook, *Hadrian's Wall Bridges*, London, 1989.

Biernacka-Labanska, M., *Mosty-Brücken im Altertum*, *Filomata,* 141, 1960.

Bihalji-Merin, Oto (ed.), *Brücken,* Lucerne, 1971.

Blake, M. E., *Ancient Roman Construction in Italy from the Prehistoric Period to Augustus*, Washington D.C., 1947.

Blake, M. E., *Roman Construction in Italy from Tiberius through the Flavians*, Washington D.C., 1959.

Blake, M. E., and D. T Bishop, *Roman Construction in Italy from Nerva through the Antonines*, Philadelphia, 1973.

Blanc, A., »Ponts gallo-romaines et très anciens d'Ardèche et de la Drôme«, *Gallia*, 25–1, 1966.

Blanco, Antonio, *El puente de Alcantara en suo context historico*, Madrid, 1971.

Blondel, Louis, »Le pont romaine de Genève«, *Bulletin de la Societé d'histoire et archéologie de Genève*, Geneva, 5, 1925–34.

Blondel, Louis, *L'emplacement du pont de César sur la Rhône*, Geneva, 1938.

Blondel, Louis, *Pont romaine de Genève*, Geneva, 1954.

Blümners, Hugo, *Technologie und Terminologie der Gewerbe und Künste bei Griechen und Römern*, 4 vols. 1874–1887, Hildesheim, 1969.

Bonatz, Paul, and Fritz Leonhardt, *Brücken*, Munich, 1956.

Bornheim, aka Schilling, »Alte und neue Brücken in Rheinland-Pfalz«, *Mittelrheinisches Jahrbuch* 2, 3, 1950/51.

Bosio, Luciano, »Ponte Sonti (Tab. Peutingeriana)«,

Borsi, Franco, *Leon Battista Alberti, The Complete Works*, New York, 1989.

Brangwyn, Frank, and Walter Shaw Sparrow, *A Book of Bridges*, London, 1915.

Brice, William Charles, *South West Asia*, London, 1966.

Briegleb, Jochen, »Die vorrömischen Steinbrücken des Altertums«, *Technikgeschichte in Einzeldarstellungen*, 14, Düsseldorf, 1971.

Briegleb, Jochen, »Review to Piero Gazzola ›Ponti Romani‹«*, Gnomon,* IXIII, 1971.

Brogan, Olwen, *Roman Gaul*, London, 1953.

Brookes, A. C., »Minturnae: The Via Appia bridge«, *American Journal of Archeology*, 78, 1974.

Brown, David J., *Bridges: Three Thousand Years of Defying Nature*, Richmond Hill, 2005.

Bruce, John Collingwood, *Handbook to the Roman Wall*, Newcastle upon Tyne, 1978.

Brussia, G., »Scavo al Ponte del Christo«, *Aquileia Nostra,* XVIII, 1947.

Buchwald, Wolfgang (ed.), et al., *Tusculum – Lexikon griechischer und lateinischer Autoren des Altertums und des Mittelalters*, Munich, 1982.

Bühler, Dirk, *Brückenbau*, Munich, 2000.

Bueno, M. A. M., *Xacimiento ibero-romano en Botorrita (Zaragoza),* XI Congreio National de Arqueologia, Merida, 1969.

Bueno, M. A. M., and J. G. M.,Valganos, *»El puento de Mantible*«, *Estudios de Arqueologia*, Alaveso, 1972.

Bundgard, J. A., *Caesar's Bridges over the Rhine*, *Acta Archaeologica*, 36, 1965/66.

Burckhardt, Jakob, *Kunst und Kultur der Renaissance in Italien*, Cologne, 1953

Burford, Alison M., »Heavy transport in classical antiquity«, *Economic History Review*, 13, 1960.

Büttner, O., and E. Hampe, *Bauwerk – Tragwerk – Tragstruktur*, Berlin, 1977.

Calder, William Moir, and George Ewart Bean, *A Classical Map of Asia Minor*, London, 1958.

Callejo Serrano, Carlos, »El puente romano de Alcantara en seco«, *Archivo Español de Arqueologia*, 43, 1970.

Calliat, Louis-Armand, »Le pont romain de Chalon«, *Memoires de la Societé d'Histoire et d'Archéologie de Chalon-sur-Saone*, XXXIII, 1952.

Carrillo, Pedro, »Via romana del summo Pyrineo à Caesar Augusta«, *Scenario do Arte Aragonés* III, Tarragona/Zaragoza, 1951.

Casado, Carlos Fernández, *Acueductos romanos en España*, Madrid, 1972.

Casado, Carlos Fernández, *Historia del Puente en España: puentes romanos*, Madrid, 1972.

Casson, Lionel, *Travel in the Ancient World*, London 1974.

Castellao, Anne-Maria, and Mauro Lotti, Il ponte dell'Arsina, *Giornale Storicao della Lunigiano,* I, 1950.

Castagnoli, Fernando, *Via appia*, Milan, 1956.

Castagnoli, Ferdinando, Antonio Maria Colini, and Giovanni Macchia, *La via appia*, Rome, 1972.

Cerutti, James, »Down the ancient appian way«, *National Geographic*, 159, 1981.

Chetto, C. S., and W. Henderson, *Masonry arch-bridges; a study*, *Institution of Civil Engineers Proceedings*, 7, 1957.

Chevallier, Raymond, *Roman Roads*, London, 1976.

Chevallier, Raymond, *Die Provinz Gallia Narbonensis*, Feldmeilen, 1979.

Chevallier, Raymond, *Voyages et déplacements dans l'empire romaine*, Paris, 1988.

Cichorius, Conrad, *Die Reliefs der Trajanssäule*, Berlin, 1900.

Clough, Arthur Hugh (ed.), *Plutarch: The Lives of the Noble Grecians and Romans /* the Dryden translation, New York, 1992.

Cohausen, August von, *Cäsars Rheinbrücken, philologisch, militärisch und technisch untersucht*, Leipzig, 1867.

Cohausen, August von, *Römische Mainbrücken*, *Mass. Ann.,* XX, 1988.

Cooke, N., »Instability of masonry arches«, *Institution of Civil Engineers Proceedings*, 83, 1987.

Coll, Pieter, *Das gab es schon im Altertum*, Würzburg, 1964.

Cornwell, Timm, J., and J. Matthews, *Atlas of the Roman World*, New York, 1982.
Cozzo, Guiseppe, *Ingegneria Romana*, Rome, 1970.
Crema, Luigi, »La vòlta nell'architettura Romana«, *L'ingegnere*, 9, 1942
Crisfield, M. A., »Finite element computer program for the analysis of masonry arches«, *Transport and Road Research Laboratory Report*, 1115.
Christ, Karl, *Die römische Kaiserzeit von Augustus bis Diokletian*, Munich, 2001.
Cüppers, Heinz, »Eine römische Straßenbrücke über die Enns«, *Bonner Jahrbücher,* 165, 1965.
Cüppers, Heinz, »Vorrömische und römische Brücken über die Mosel«, *Germania*, 45, 1967.
Cüppers, Heinz, »Die Trierer Römerbrücken«, *Trierer Grabungen und Forschungen,* vol. V, Mainz, 1969.
Curtius, Ludwig, *Das antike Rom*, Vienna, 1943.
Curtius, Ludwig, *Torso, verstreute und nachgelassene Schriften*, Stuttgart, 1957.

Dantremont, Leon, »*Les sept pons brivois*«, *Bulletin de la Société scientifique de la Corrèze*, IXXX, 1959.
De Angelis d'Ossat, G., »Architettura romana a vòlta, 1935–1940«, *Bull. Arch. Com.,* 1940.
De Camp, Lyon Sprague, *The Ancient Engineers*, New York, 1963.
Decci, A., »Le pont de Traian à Turnu-Severin, *Ann. Inst. De studis classice*«, 1928/29.
Degraassi, N., »*Il* ponte romano di Olginato e la strada da Bergamo a Como«, *Rivista Archeologica dell'antica Provincia e Diocesi di Como*, CXXVII, 1946.
Deichmann, Friedrich Wilhelm, *Westliche Bautechnik im römischen und rhomäischen Osten*, *Mitteilungen des Deutschen Archäologischen Instituts*, *Römische Abteilung*, vol. 86, Mainz, 1979.
Deman, Esther Boise van, *The building of Roman Aqueducts*, Washington, 1934.
Diels, Hermann, *Antike Technik*, Leipzig, 1924.
Dietrich, Richard J., *Faszination Brücken*, Munich, 1998.
Dilke, Oswald Ashton Wentworth, *The Roman Land Surveyors: An Introduction to the Agrimensores*, Newtown Abbot, 1971.
Dilke, Oswald Ashton Wentworth, *Greek and Roman Maps*, London, 1985.
Dischinger, F., *Massivbau*, *Taschenbuch für Bauingenieure*, Berlin, 1949.
Donini, Augusto, *Ponti sur moneto e medaglie vicende construttive*, *Noticie storico Roma*, Santamarina, 1959.
Döring, Mathias, »Die römische Wasserleitung von Pondel (Aostatal)«, *Antike Welt*, 29, 2, 1998.
Dörner, Friedrich Karl (ed.), *Vom Bosporus zum Ararat*, Mainz, 1984.
Dörner, Friedrich Karl, *Der Thron der Götter auf dem Nemrud Dâg*, Bergisch Gladbach, 1987.
Dörpfeld, Wilhelm (ed.), *Troia und Ilion*. *Ergebnisse der Ausgrabungen in den vorhistorischen und historischen Schichten von Ilion 1870–1894*, vol.1, Osnabrück, 1968.
Duddeck, Heinz, »Entwicklung der Berechnungsmodelle des Bauingenieurs: Woher? Wohin?«, *Bautechnik,* 70, 1993.
Dufournet, Paul, »Ponts et passages du Rhône entre, Le Pas-de-L'Ecluse et Yenne, et le réseau routier correspondant dans l'Antiquité«, *La Revue Savoisienne,* CXIII, 1973.
Durm, Josef, *Die Baukunst der Etrusker und die Baukunst der Römer*, *Handbuch der Architektur*, Stuttgart, 1905.
Dyggve, Ejnar, »Archäologie und Statik«, *Österreichisches Jahrbuch*, 43, 1956.
Dymond, David Percy, »Roman bridges on Dere Street, County Durham«, *Archaeological Journal*, 118, 1961.

Emerson, William, *Old Bridges of France, a Series of Historical Examples from Roman Times to the End of the XVIIIth Century*, New York, 1925.
Engels, Michael, Matthias Hupperich, Ralf Müller, and *Michael* Olbering, *Die Wasserleitung nach Side, eine Betrachtung aus bautechnischer Sicht*, (thesis), Technical University Aachen, 1984.
Enziclopèdia dell'Arte Antica, Classica e Orientale, entry »Ponte«, Rome, 1958–66.
Erler, Uwe, and Helga Schmiedel, *Brücken*, Leipzig 1988.
Ertman, P.C., *Curatores Viarum, a Study of the Superintendents of Highways in Ancient Rome*, Ann Arbor, 1976.
Esch, Arnold, »Die Via Cassia in der Landschaft«, *Antike Welt*, 19, 1, 1988.
Esch, Arnold, »Die Via Flaminia in der Landschaft«, *Antike Welt*, 26, 2, 1995.
Esch, Arnold, *Römische Straßen in ihrer Landschaft*, Mainz, 1997.
Eschebach, Hans, »Pompeji – Straßenbau in der Antike«, *Antike Welt*, 9, 4, 1978.
Essen, C. C. van, »The Via Valeria from Tivoli to Collarmele«, *Papers of the British School at Rome*, 25, 1957.

Fehr, Horst, »Eine Rheinbrücke zwischen Koblenz und Ehrenbreitstein aus der Zeit des Claudius«, *Bonner Jahrbücher,* 181, 1981.
Feldhaus, Franz Maria, *Die Technik der Antike und des Mittelalters*, Potsdam, 1931.
Fentress, Elisabeth, »Via Aurelia, Via Aemilia«, *Papers of the British School at Rome*, 52, 1984.
Fiches, Jean-Louis*, Un ouvrage d'art sur la voie Domitienne, le pont d'Ambrussium*, *R. St. Lig*. XXXVI, Montpellier, 1970.
Fiches, Jean-Louis, »L'oppidum d'Ambrussum, le pont romain, le quartier bas«, *A.R.A.L.O.*, Caveirac ,1982.
Finley, Moses J., *The Ancient Economy*, Berkeley, 1973.
Forbes, Robert James, *Notes on the History of Ancient Roads and their Construction*, Amsterdam, 1934.
Forbes, Robert James, *Studies in Ancient Technology*, Leiden, 1964–71.
Forbes, Robert James, *Ancient Roads*, Leiden, 1964.
Formigé, Jules, *Les monuments romains de la Provence*, Paris, 1974.
Frank, T., *Roman Buildings of the Republic, an attempt to date them from their materials*, Rome, 1924.
Frontinus, Sextus Julius, *De aquia urbis Romae. The Two Books on the Water Supply of the City of Rome of Sextus Julius Frontinus*, *Water Commissioner of the City of Rome A. D. 97*. A translation into English, and explanatory chapters by Clemens Herschel, New England Water Works Association, 1973 (2010).
Frontinus Gesellschaft e.V., *Die Wasserversorgung antiker Städte*, vols. 2 and 3, Mayence, 1987/88.
Froriep, Siegfried, »Ein Wasserweg in Bythnien«, *Antike Welt*, 17, 2, 1986.
Fustier, Pierre, *La route. Voies antiques, chemins anciens, chaussées modernes*, Paris, 1968.

Gaitzsch, W., *Eiserne römische Werkzeuge*, *British Archaelogical Reports, international series*, 2 vols., Oxford, 1980.
Gajewska, Halina, »The Roman bridges and harbours on the lower Danube«, *Archeologia*, XX, 1969.
Gall, Jòel, *Le Tibre, fleuve de Roma*, *l'Antiquité*, Paris, 1952.
Galli, Edoardo, and Giuseppe Viccei, *Restauro del ponte romano di Porta Cappucina in Ascoli Piceno,* Istituto d'Arte del Libro (no place), 1940.
Galliazzo, Vittorio, *Il ponte della Pietra di Verona*, Verona, 1969/70.
Galliazzo, Vittorio, *Il ponti di Padova romana*, Padova, 1971.
Garcia y Bellide, Antonio, »El puente romano de Medellin«, *Archivo Espanol de Arqueologia Madrid,* XXVI, 2, 1953.
Gavelle, E., »Aux abords de saint-Bertrand-de-Comminges«, *Revue de Comminges*, XXIV, 1961.
Gazzola, Pietro, *Ponti Romani*, Florence, 1963.
Gazzola, Pietro, *Ponte Pietra a Verona*, Florence, 1964.
Germain de Montanzan, C., *La science de l'art de l'ingenieur aux premiers siècles del'empire romain*, Paris, 1909.
Gibbon, Edward, *The Rise and Fall of the Roman Empire*, London, 1892
Gilles, R. C., »How Caesar bridged the Rhine«, *Classical Journal*, 64, 1968/69.
Giovannoni, G.,»La technica delle costruzioni romane a vòlta«, *Atti Società It. Progresso Scienze*, XVIII, Florence, 1929.
Gose, E., »Neue Beobachtungen an der Römerbrücke«, *Trierer Zeitschrift,* 27, 1964.
Graefe, Rainer, *Zur Geschichte des Konstruierens*, Stuttgart, 1989.
Grant, Michael, *Ancient History Atlas*, London, 1978.
Greene, Kevin, »Perspectives on Roman technology«, *Oxford Journal of Architecture,* 9, Oxford, 1990.
Grenier, Albert, *Manuel d'archéologie gallo-romaine*, 4 vols., Paris, 1931.
Grewe, Klaus, »Straßen, Brücken und Meilensteine, Spuren römischer Infrastruktur in England«, *Antike Welt,* 26, 5, 1995.
Grewe, Klaus, *Planung und Trassierung römischer Wasserleitungen*, Wiesbaden, 1985.
Grewe, Özis, Baykan, and Atalay, »Die antiken Flussüberbauten von Pergamon und Nysa (Turkey)«*, Antike Welt*, 25, 4, 1994.
Grewe, Klaus, *Licht am Ende des Tunnels: Planung und Trassierung im antiken Tunnelbau*, Mainz, 1998.
Grewe, Klaus, *Großbritannien, Ein Führer zu bau- und technikgeschichtlichen Denkmälern aus Antike und Mittelalter*, Stuttgart, 1999.
Grewe, Klaus, Paul Kessener, and Susanna Piras, »Im Zickzack-Kurs über den Fluss: Die römisch-

seldschukische Eurymedon Brücke von Aspendos (Turkey)«, *Antike Welt*, 30, 1999.
Gündel, Friedrich Herrmann, »Die römische Mainbrücke bei Frankfurt a.M.«, *Germania*, 6, 1922.
Günther, Adam, »Römische Rheinbrücke zwischen Koblenz und Ehrenbreitstein«, *Bonner Jahrbücher,* 146, 1941.
Guzzo, Piero G., »Sacrofano, ponte romano in località Fontana Nuova«, *Notizie degli Scavi di antichità*, XXIV, 1970.

Hagen, Josef, »Römerstraßen in der Rheinprovinz«, *Erläuterungen zum geschichtlichen Atlas der Rheinprovinz*, VIII, 1931.
Hagen, Victor Wolfgang von, *The Roads that led to Rome*, London, 1967.
Hammond, N. G. L., »The western part of the Via Egnetia«, *Journal of Roman Studies*, 64, 1974.
Handbuch der Ingenieurwissenschaften, vol. 2 (*Der Brückenbau*), Leipzig, 1906.
Hart, Fritz, *Kunst und Technik der Wölbung*, Munich, 1965.
Harvey, W. J., F. W. Smith, and A. E. Vardy, »Analysis and design of masonry arches«, *Proceedings of the International Symposium on Bridges and Structures*, Lanzhou (China), 1978.
Haselberger, Lothar, »Antike Bauzeichnungen des Pantheon«, *Antike Welt*, 25, 4, 1994.
Haselberger, Lothar, »Die Werkzeichnungen des Naiskos im Apollotempel von Didyma«, *Diskussion zur archäologischen Bauforschung*, 4, Berlin, 1984.
Haselberger, Lothar, »The construction plans of the temple of Apollo at Didyma«, *Scientific American*, Dec. 1985.
Hasluck, Frederick William, »A Roman bridge on the Aesepos (Mysia)«, *Annals of the British School of Athens,* 12, 1905.
Hauck, C. F., »Structural design of the Pont du Gard«, *Journal of American Society of Civil Engineers,* 112, 1986.
Hausen, K., and R. Rürup (ed.), *Moderne Technikgeschichte*, Cologne, 1975.
Haverfield, C., *Roman Roads in Ancient Britain*, Oxford, 1939.
Heimberg, Ursula, *Römische Landvermessung*, *Kleine Schriften zur Kenntnis der römischen Besatzungsgeschichte Südwestdeutschlands,* 17, Stuttgart, 1977.
Heinrich, Bert, *Brücken: Vom Balken zum Bogen*, Reinbek, 1983.
Heinz, Werner, »Straßen und Brücken im Römischen Reich«, *Antike Welt*, *Sondernummer*, 2, 1988.
Heyden, A. A. M. van der, and H. H. Scullard, *Atlas of the Classical World*, London, 1959.
Heymann, Jacques, »The stone skeleton. Structural Engineering of Masonry Architecture«, *International Journal of Solids and Structures*, 2, issue 2, 1966.
Heymann, Jacques, »The safety of masonry arches«, *International Journal of Mechanical Science*, 11, issue 4, 1969.
Heymann, Jacques, *Plastic Design of Frames*, vol. 2, *Applications*, Cambridge, 1971.
Heymann, Jacques, *Coulomb's Memoir on Statics, an Essay in the History of Civil Engineering*, Cambridge, 1972.
Heymann, Jacques, *Couplet's Engineering Memoirs 1726–33*, Cambridge, 1972.
Heymann, Jacques, »Estimation of strength of masonry arches«, *Proc. Instn Civ. Engrs.*, 69, 1980.
Heymann, Jacques, *The Masonry Arch*, Cichester, 1982.
Heymann, Jacques, and C. J. Padfield, »Two masonry bridges. I. Clare College Bridge«, *Proc. Instn Civ. Engrs.*, 52, 1972.
Hill, Donald Routledge, *A History of Engineering in Classical and Medieval Times*, London, 1984.
Hirschfeld, Kurt, *Baustatik I and II*, Heidelberg, 1969.
Horace (Quintus Horatius Flaccus), *Satires and Epistles of Horace*, translated by John Conington, Oxford, 2008.
Höper, H., »Die Brücke von Xerokambios (Greece)«, *Boreas*, 4, 1931.
Hutchinson, A. W., »A Roman Bridge in the Fens«, *Ant. J.* 14, 1934.

Jacopi, Giulio, »Il ponte romano di Kiahta«, *Bolletino d'Arte*, XIX, 1936/37.
Jennings, A., »Stability fundamentals in relation to masonry arches«, *Struct. Eng.* B64, 1986.
Jordan, H., *Topographie der Stadt Rom im Altertum*, vol. 2, London, 1929.
Jurecka, Charlotte, *Brücken, historische Entwicklung, Faszination der Technik*, Vienna, 1979.
Jüngst, E., and P. Thielscher, »Vitruv über Baugrube, Baugrund und Grundbau, ein Interpretationsversuch«, *Römische Mitteilungen*, 51, 1936.

Kammüller, K., and W. Swida, *Statische Untersuchung eines fest eingespannten Brückengewölbes, Karlsruhe,* 1948.
Kahane, A., »The paved Roman road east from Gabii«, *Papers of the British School at Rome*, 41, 1975.
Keller, Werner, *Denn sie entzündeten das Licht. Geschichte der Etrusker*, Munich, 1970.
Kelsch, Wolfgang (ed.), *Rom in alten Stichen*, Hanover, 1975.
Kenny, E. J. A., »A Roman bridge in the Fens«, *Geographic Journal*, 82, 5, 1933.
Kittel, Th., and O. Stoll, »Die Brücke bei Nimreh. Ein Zeugnis römischer Verkehrspolitik im Hauran (Syria)«, *Antike Welt*, 31, 2.
Klein, Michael, »Untersuchungen zu den kaiserlichen Steinbrüchen an Mons Porphyrites und Mons Claudius in der östlichen Wüste Ägyptens«, *Alte Geschichte,* 26, 1988.
Klugelhofer, H., *Römische Technik*, Zurich, 1951.
Kissel, Theodor, »Veliti naturae ipsius dominus«, *Antike Welt*, 33, 2.
Knapp, R. C., *Roman Corduba*, Berkeley, 1983.
Köthe, Rainer, *Brücken*, Nuremburg, 1991.
Kottmann, Albrecht, *Bogen und Gewölbe*, Munich, 1974.
Krause, Rüdiger, »Eine römische Straßenstation an der Nordgrenze der Provinz Rätien«, *Antike Welt*, 24, 2.
Krafft, F., *Dynamische und statische Betrachtungsweise in der antiken Mechanik*, *Gymnasium,* 81, 1970.
Kretschmer, Fritz, *Technik und Handwerk im Imperium Romanum*, Düsseldorf, 1983.
Kretschmer, Fritz, *Bilddokumente römischer Technik*, Düsseldorf, 1983.
Korres, M., »Vorfertigung und Ferntransport eines athenischen Großbaus. Proportionierung von Säulen in der hellenistischen Architektur«, *Bauplanung und Bautheorie der Antike*, *Diskussionen zur Archäologischen Bauforschung*, 4, Kolloquium, Berlin-West, 1983.
Küffner, Georg, »Brücken sind Weg, Zeit und Ziel«, *Frankfurter Allgemeine Zeitung*, 31.12.2001.

Lamas, Manuel, »El puente romano de Orense«, *Cuadernos de Estudios Gallegos*, XI, 1956.
Lamboglia, Nino, »Cinque ponti, Val Ponci, Finale Liguria«, *Notiziario Archeologico, Rivista Luguana e Intermelia*, IX, 1954.
Lamboglia, Nino, »Un monumento defunto: il pontetto romano de Loano«, *Revista Luganua e Intermelia*, XV, 1960.
Landels, John G., *Engineering in the Ancient World*, London, 1978.
Lamprecht, Heinz Otto, *Opus caementitium, Bautechnik der Römer*, Düsseldorf, 1984.
Leonhardt, Fritz, *Brücken, Ästhetik und Gestaltung*, Stuttgart, 1982.
Lepper, F., and S. S. Frere, *Trajan's Column. A New Edition of the Cichorius Plates*, Gloucester, 1988.
Levi, D., *Le grande strade romane in Asia*, Rome, 1938.
Livius, Titus, *Ab urbe condita*, translated by John Henry Freese, Alfred John Church, and William Jackson Brodribb, with a critical and biographi cal introduction and notes by Duffield Osborne, New York, 1898.
Llopis, Salvador, »*Restos de un Puente romano sobre el rio Bernisa, Gandia*«, *Cronica de VI. Congresso Arqueologia*, Cartagena, 1951.
Lugli, Guiseppe, *Forma Italiae*, pars I, Rome, 1926; pars II, Rome, 1928.
Lugli, Guiseppe, *I monumenti antichi di Roma e suburbio*, 4 vols., Rome, 1930–38, suppl. 1940.
Lugli, Guiseppe, *Roma antica*, Rome, 1946.
Lugli, Guiseppe, *Consideratione sull'origine dell'arco a conci radiale*, Rome, 1952.
Lugli, Guiseppe, *La technica edilizia romana con particolare riguardo a Roma e Lazio*, Rome, 1957.

MacPherson, I. W., »Roman roads and milestones of Galatia«, *Anatolian Studies*, 4, 1954.
Manière, Gabriel, »Le pont du Diable dans la commune de Saint-Christand est d'origine romaine«, *Revue de Comminges*, IXXX, 1967.
Manière, Gabriel, »Voies et ponts antiques dans la commune de Saint-Marlory (Haute Garonne)«, *Gallia,* XXIII, 1969.
Martinori, F., *Via Flaminia*, Rome, 1929.
Mauer, F., »Die alte Römerbrücke in Heilbronn-Neckargartach«, *Schwäbische Heimat*, 12, 1961.
Mayerhöfer, A., *Die Brücken im alten Rom*, Erlangen, 1883.
Meighörner-Schadt, Wolfgang, and Heinrich Blumenthal, *Räderwerk Römerkran*, Bonn, 1989.
Meißner, Burkhard, *Die technologische Fachliteratur der Antike. Struktur, Überlieferung und Wirkung technischen Wissens in der Antike*, Berlin, 1999.
Mensching, Eckart, »P. Pomponius Secundus und der Brückenbau an Rhein und Mosel: Die Koblenzer Rheinbrücke«, *Bonner Jahrbücher,* 181, 1981.

Melan, Joseph, *Der Brückenbau*, vol. 2, Leipzig/Vienna, 1924.
Merckel, Curt, *Die Ingenieurtechnik im Altertum*, Berlin, 1902.
Merino, Garcia Carmen, »*El puente romano inèdito de Golmayo (Soria)*«, *B. em. EAA-Valladolid*, XXXIX, 1973.
Mesqui, Jean, *Le pont en France avant le temps des ingénieurs*, Paris, 1986.
Meyer, W., »Die Römerbrücke bei Kroisegg (Austria)«, *Burgenländische Heimatblätter*, 37, 1975.
Mitford, T.B., »Some inscriptions from the Cappadocian Limes«, *Journal of Roman Studies*, 64, 1974.
Mommsen, Theodor, *Römisches Staatsrecht,* vol. 3, part 1, Berlin, 1899.
Monkcwitz, Kurt, »*Der Pont Julien, ein römisches Bauwerk im Herzen der Provence*«, *Antike Welt*, 13, 4, 1982.
Mörsch, E., *Statik der Gewölbe und Rahmen*, part 1, Stuttgart, 1947.
Müller, W., *Architekten in der Welt der Antike*, Munich, 1989.
Müller-Breslau, Heinrich Franz, Die *graphische Statik der Baukonstruktionen*, vol. II, Leipzig, 1925.
Müller, Werner, and Gunther Vogel, *DTV-Atlas zur Baukunst*, vol. 1, Munich, 1977.
Müller-Wiener, Wolfgang, *Griechisches Bauwesen in der Antike*, Munich, 1988.
Muurinho, Antonio M., »Ponte romana no rio Tuela e sintese das vias e pontes romanas do nordeste transmontano«, *Trabalhos de Antropologia et Etnologia,* XXIII, 1978.

Nash, Ernest, *Pictorial Dictionary of Ancient Rome*, 2 vols., London, 1968.
Neuburger, Albert, *Die Technik des Altertums*, Leipzig, 1929, 1987.
Nikolajev, J.S., »La technique antique à la construction du Pont du Gard«, *Sov. Arch*. 2, 1967.
Niessen, Heinrich, *Italienische Landeskunde*, Berlin, 1902.
Noaek, F., *Die Baukunst des Altertums*, Berlin, 1912.
Norberg-Schulz, Christian, *Intentions in Architecture*, Oslo, 1963.

O'Connor, Colin, *Roman Bridges*, Cambridge, 1993.
Olshausen, Eckart, *Einführung in die Historische Geographie der Alten Welt*, Darmstadt, 1991.

Page, J., »Load tests to collapse on two arch bridges at Preston Shropshire and Prestwood Staffordshire«, Research Report, 110, 1987
Palladio, Andrea, *The Four Books on Architecture*, Dover, 1977.
Pekary, Thomas, *Untersuchungen zu den römischen Reichsstraßen*, *Antiquitas, Reihe 1*, Bonn, 1968.
Perret, Vincent, »Le pont vetus de Narbonne«, *Bulletin da la Commission archéologique de Narbonne*, XXXI, 1969.
Persichetti, Niccolò, *La Via Salaria nel circondario di Ascoli Piceno*, Rome, 1904.
Petersen, Christian, *Statik und Stabilität der Baukonstruktionen*, Wiesbaden, 1982.
Picard, Gilbert, *Imperium Romanum. Architektur der Welt*, Fribourg, 1965.

Pippard, A.J.S., E. Tranter, and L. Chitty, »The mechanics of the voussoir arch«, *Institution of Civil Engineers Journal*, 4, 1936.
Platner, S.B., and T.S. Ashby, *A topographical Dictionary of Ancient Rome*, Oxford, 1929.
Plinius the Younger, *C. Plini Caecili Secundi: Epistularum libri decem*, Oxford, 1963.
Poellnitz, P., *Die römische Rheinbrücke bei Mainz*, Mainz, 1884.
Potter, Timothy W., *Roman Italy*, London, 1987.
Prade, Marcel, *Les Ponts – Monuments Historiques*, Poitiers, 1986.
Princeton Encyclopedia of Classical Sites (ed.), Princeton, 1976.
Prokopius of Kaisareia: *De aedificiis (Building under Justinian)*, Stuttgart, 1962–64.

Quilici, Gigli S., La *Via Salaria da Roma a Passo Corese*, Rome, 1977.
Quilici, Lorenzo, *Via Appia*, Rome, 1989.
Quilici, Lorenzo, *La Via Prestina, i suoi monumenti, i suoi paesaggi*, *Collection Italia nostra*, Rome, 1977.
Quilici, Lorenzo, *La Via Latina da Roma a Castel Savelli*, *Collection Italia nostra*, Rome, 1978.
Quilici, Lorenzo, *La Via Appia dalla Pianura Pontina a Brindisi*, Rome, 1989.
Quilici, Lorenzo, G.M. De Rossi, and P. G. Dominico, »La Via Aurelia da Roma a Civitavecchia«, *Quaderni dell'Istituto di topografia dell'Universita di Roma*, IV, Rome, 1968.

Radke, Gerhard, *Viae Publicae Romanae*, *Realencyclopädie der classischen Altertumswissenschaft*, supplement 13, Stuttgart, 1971.
Rakop, F., *Römische Architektur in Nordafrika, Bautechnik und Bautradition, Report on 150th anniversary of German archeological Institute*, Rome, 1982
Ramme, Werner, *Über die geschichtliche Entwicklung der Statik im Bauwesen*, Glentorf, 1939.
Raven, Susan, *Rome in Africa*, London, 1969.
Reindl, C., *Brücken und Pegel im Altertum*, Wasserkraft und Wasserwirtschaft, 35, 1940.
Reuther, Hans, *Römische Wölbkunst*, *Antike Welt*, 3, 1971.
Richmond, I. A., *The aqueducts of ancient Rome*, Oxford, 1935.
Richmond, Ian A., *Roman Britain*, London, 1947.
Rivet, Albert Lionel Frederick, »Gallia Narbonensis, with a Chapter on Alpes Maritimae«, *Southern France in Roman Times*, London, 1988.
Rivoira, G.T., *Roman Architecture and its principles of construction under the Empire: with an appendix on the evolution of the dome up to the XVIIth century*, translated from the Italian by G. McN. Rushfort, 1925, reprinted New York, 1972.
Robertson, A., »River Crossings on the Antonine Wall«, *Roman Frontier Studies*, Cardiff, 1969.
Robertson, Donald Struan, »Construction Techniques for Arches, Vaults and Domes«, *Handbook of Greek and Roman Architecture*, Cambridge, 1943.
Robins, Frederick William, *The Story of Water Supply*, Oxford, 1946.
Robins, Frederick William, *The Story of the Bridge*, Birmingham, 1948.

Röder, Georg Ludwig Alexander, *Praktische Darstellung der Brückenbaukunde nach ihrem ganzen Umfange. Nach den bewährtesten Technikern und Mathematikern und den besten vorhandenen Mustern jeder Art. Erster Theil: Hülfskenntnisse und den Bau steinerner Brücken enthaltend*, *Zweyter Theil: den Bau der hölzernen, eisernen und beweglichen, sowie der Nothbrücken enthaltend*, Darmstadt, 1845/46.
Rösger, Alfons, and Wolfgang Will, Die Drususbrücke zu Bonn, *Bonner Jahrbücher,* 185, 1985.
Rostovtzeff, Michael, I., *The Social and Economic History of the Roman Empire*, Oxford, 1957.
Rupp, Erwin, *Bautechnik im Altertum*, Munich, 1964.

Sadler, Heiner, *Brücken*, Dortmund, 1987.
Salama, Pierre, *Les voies romain de l'Afrique du Nord*, Paris, 1951.
Scarfi, B.M., »Colfoso, ponte romana sulla Via Claudia Augusta«, *Bolletino d'Arte*, XIX, 4, 1964.
Selberg, A., »On the bearing capacity of voussoir arches«, *IABSE*, 13, 1953.
Selkirk, Raymond, *The Piercebridge Formula*, Cambridge, 1983.
Sevillano, Virgilio, »La calcada de Miranda en Zamora y sus puentes«, *B.Sem. EAA*, XXXVI, 1977.
Singer, C. (ed.), *Oxford History in Ancient Technology*, Oxford, 1956.
Sitwell, Nigel H.H., *Roman Roads of Europe*, London, 1981.
Soeder, Hans, *Urformen der abendländischen Baukunst in Italien und im Alpenraum*, Cologne, 1964.
Spratt, Thomas Abel Brimage, and Edward Forbes, *Travels in Lycia, Mylias, and the Cibyratis, in Company with the late Rev. E. T. Daniell*, London, 1847.
Starr, S. Frederick, »Mapping ancient Roads in Anatolia«, *Archeology*, XVI, 1963.
Stenico, Arturo, »Trovamenti archeologico no letto de Ticino a Pavia«, *Ticino*, VI, 1950.
Sterpos, D., »La strada romana in Italia«, *Autostrada*, 17, Rome, 1970.
Strabo, *The Geography of Strabo*, transl. and ed. by Horace Leonard Jones and John Robert Sitlington Sterrett, London 1917–1932.
Straub, Hans, *Geschichte der Bauingenieurkunst*, Basel, 1992.
Swida, Woldemar, *Statik der Bogen und Gewölbe,* Karlsruhe, 1954.
Szabó, István, *Geschichte der mechanischen Prinzipien und ihre wichtigsten Anwendungen*, Basel, 1976.

Schetter, W., and R. von Ustar, »Zu den pontes longi«, *Gymnasium*, 87, 1971.
Schiefendecker, Franz-Dietrich, »Brückentechnische Überlegungen zur römerzeitlichen Rheinbrücke bei Koblenz«, *Bonner Jahrbücher*, 181, 1981.
Schmidt, Fritz, *Geschichte der geodätischen Instrumente und Verfahren im Altertum und Mittelalter*, Neustadt a.d.H., 1935
Schreiber, Georg, *Die Römer in Österreich*, Frankfurt am Main, 1974.
Schreiber, Hermann, *Auf Römerstraßen durch Europa*, Munich, 1985.
Schumacher, Leonhard (ed.), *Römische Inschriften*, Stuttgart, 1988.

Schumacher, R., »Die Drusenbrücke über die Nahe bei Bingen«, *Jahrbuch für Geschichte und Kunst des Mittelrheins*, 1952/53.

Tabula Peutingeriana, facsimile reproduction, Graz 1976.
Talbert, Richard J. A. (ed.), *Atlas of Classical History*, London, 1985.
Talip, Semseddin, *Le strade romane in Anatolia*, Rome, 1938.
Taradelli, Miguel, »Investigaciones arqueológico en la provincia de Granada«, *Ampurias*, IX–X, 1947/48.
Texier, Charles, *Description de l'Asie Mineure I* (*Bridge over the Sarkarya at Adapazari*), Paris, 1839/40.
Tomassetti, Giuseppe, *La Campagna Romana antica, mediovale e moderna*, Rome, 1910.
Torroja, Eduardo, *Philosophy of Structures*, Los Angeles, 1958.
Tozzi, Pierluigi, »Il ponte romano di Pavia«, *Annali Benacensi*, 7, 1981.
Travlos, John, »Hadrian's Bridge on the Eleusinian Kephissos«, *American Journal of Archeology,* IV, 1951.
Tudor, D., »Un pont romain ignoré dans la region du Bas-Danube«, *Latomus*, XX, 3, 1961.
Tudor, D., *Un pont romain du Bas-Danube*, Bucarest, 1974.
Tunc, Gülgün, *Tas Köprülerimiz*, Ankara, 1978.
Tyrrell, Henry Grattan, *History of Bridge Engineering*, Chicago, 1911.

Vilnay, O.,»Buckling of masonry arches«, *Institution of Civil Engineers Proceedings*, 77, 1984.
Vitruvius, (Marcus Vitruvius Pollio), *De architectural ibri decem, Rome, 1960*
Vitruvius*, The Ten Books on Architecture,* London, 1826.
Vinson, S.P., »Roman Piers on the Via Herculia«, *Archaeology*, XXIX, 1976.

Wacher, John (ed.), *The Roman World*, London, New York, 1987.
Wagner, Jörg, »Die Römer am Euphrat«, *Antike Welt*, 6, 1975.
Walklate, R. P., »Method of determining the permissible loading of brick and masonry arches«, *Institution of Civil Engineers Proceedings*, 75,1983.
Ward-Perkins and John Bryan, *Roman Architecture,* London, 2003.
Ward-Perkins, John Bryan, »Etruscan and Roman roads in southern Etruria«, *Journal of Roman Studies*, London, 57, 1957.
Ward-Perkins, John Bryan, »Quarrying in Antiquity – Technology, Tradition and Social Change«, *Proceedings of the British Academy*, 57, London, 1973.
Walser, Gerold, »Die römischen Straßen und Meilensteine in Rätien«*, Kleine Schriften zur Kenntnis der römischen Besatzungsgeschichte*, 29, Stuttgart, 1983.
Weber, Ekkehard, *Die Tabula Peutingeriana,* Graz, 1976, and *Antike Welt,* 15, 1984.
Whitby, L. M., »Justinian's bridge over the Sangarius and the date of Procopius' De aedificiis«, *Journal of Hellenic Studies*, 105, 1985.
White, K. D., *Greek and Roman Technology. Aspects of Greek and Roman Life*, London, 1984.
Wiseman, F. J., *Roman Spain, an Introduction to the Roman Antiquities of Spain and Portugal*, London, 1956.
Wiseman, Timothy Peter, »Viae Anniae«, *Papers of the British School at Rome*, 32, 1964.
Wiseman, Timothy Peter, »Viae Anniae again«, *Papers of the British School at Rome,* 37, 1969.
Wiseman, Timothy Peter, »Roman Republican Road Building«, *Papers of the British School at Rome*, 38, 1970.
Winbolt, S. E., »Remains of the Roman Bridge over the Arun (Sussex)«*, Archaeological Collections*, 76, 1935.
Winslow, Earle Micajah, *A Libation to the Gods, the Story of Roman Aqueducts*, London, 1963.
Wölfel, Wilhelm, *Wasserbau in den Alten Reichen*, Berlin, 1990.
Wurster, Wolfgang, and Joachim Ganzert, *Eine Brücke bei Limyra in Lykien*, *Archäologischer Anzeiger*, Berlin, 1978.

Zannini, Gian Ludovico, »Restauri al ponte di Tiberio nel sec. XIX«, *Rimini, Storia, Arte e Cultura*, 1969.
Ziesemer, Helmut, *Gebrauchsfertige Formeln für den gelenklosen Kreisbogen*, *Beton- und Stahlbetonbau*, 47, 1952.
Zimmerhaeckel, F., »Caesars Rheinbrücke«, *Zeitschrift für mathematischen und naturwissenschaftlichen Unterricht*, Leipzig, 1899.
Zucker, Paul, *Die Brücke, Typologie und Geschichte ihrer künstlerischen Gestaltung*, Berlin,1921.